Advances on Fine Particles and Bubbles Flotation

Advances on Fine Particles and Bubbles Flotation

Editors

Liuyi Ren
Wencheng Xia
Wei Xiao
Siyuan Yang

MDPI

Basel • Beijing • Wuhan • Barcelona • Belgrade • Novi Sad • Cluj • Manchester

Editors

Liuyi Ren
School of Resources and
Environmental Engineering
Wuhan University
of Technology
Wuhan
China

Wencheng Xia
School of Chemical
Engineering and Technology
China University of Mining
and Technology
Xuzhou
China

Wei Xiao
School of Resources
Engineering
Xi'an University of
Architecture and Technology
Xi'an
China

Siyuan Yang
School of Resources and
Environmental Engineering
Wuhan University
of Technology
Wuhan
China

Editorial Office
MDPI
St. Alban-Anlage 66
4052 Basel, Switzerland

This is a reprint of articles from the Special Issue published online in the open access journal *Minerals* (ISSN 2075-163X) (available at: www.mdpi.com/journal/minerals/special_issues/AFPBF).

For citation purposes, cite each article independently as indicated on the article page online and as indicated below:

Lastname, A.A.; Lastname, B.B. Article Title. *Journal Name* **Year**, *Volume Number*, Page Range.

ISBN 978-3-7258-1136-6 (Hbk)
ISBN 978-3-7258-1135-9 (PDF)
doi.org/10.3390/books978-3-7258-1135-9

Contents

About the Editors

Liuyi Ren

Liuyi Ren is a Professor at the School of Resources and Environmental Engineering, Wuhan University of Technology, China. Her research interests include fine particle flotation, micro bubble flotation, selective grinding, interfacial chemistry, solid waste recycling, etc. She obtained her Ph.D. degree in Mineral Processing Engineering from Central South University, China in 2012, and has worked in WUT since then. She visited the School of Chemical Engineering, University of Queensland, Australia as a visiting scholar in 2017. She obtained three projects funded by the National Natural Science Foundation of China, three projects funded by the China Hubei Provincial Science Technology Department, and many other projects funded by Social Society. She has published more than 60 papers in well-known journals such as *Minerals Engineering*, and has applied for more than 10 patents. She obtained three provincial and ministerial science and technology awards, and three teaching achievement awards, etc. She obtained the second prize of the 11th Young Teachers' Teaching Competition of Wuhan University of Technology. She is the youth Editorial Board Member of *Nonferrous Metals* and the Guest Editor of *Minerals*.

Wencheng Xia

Wencheng Xia, Ph.D., is an Associate Professor at China University of Mining and Technology (CUMT). He obtained his Ph.D. in 2014. His research interests are mineral flotation, fine particle dewatering, surface modification of coal/minerals, etc. He obtained two projects funded by the National Natural Science Foundation of China, and other projects funded by Social Society, industrial plants, etc. He has published more than 150 papers in intentional journals and acts as a reviewer for journals such as *Fuel, Energy Fuels, Powder Technology, Minerals Engineering*, etc. He was awarded the Highly Cited Chinese Researchers award (2020, 2021, 2022, 2023).

Wei Xiao

Wei Xiao is an Associate Professor and master's supervisor at Xi'an University of Architecture and Technology, and a candidate for the "Kunlun Talents" program in Qinghai Province. He is the Youth Director of the Chinese Particle Society and Member of the Micro Nanobubble Professional Committee of the Chinese Particle Society; Young Editorial Board Member of the *Journal of Central South University* (Natural Science Edition) and *International Journal of Mines, Metallurgy and Materials*. He mainly engaged in research on the development of green flotation reagents, as well as the theory and technical application of nanobubbles. He has hosted six national- and provincial-level projects, including the National Natural Science Foundation of China, China Postdoctoral Science Foundation, and Shaanxi Provincial Natural Science Foundation, as well as three horizontal projects commissioned by enterprises and institutions. He has published over 40 SCI papers in international journals such as *ME*, 1 academic monograph, authorized 5 national invention patents, 5 utility model patents, authorized 15 software copyrights, participated in writing 2 national standards, and 1 group standard. In 2021 and 2023, he won second prize of the Shaanxi Province Higher Education Science and Technology Progress Award and one innovation-driven award (both ranked first), and first prize of the 2023 Non-Metallic Science and Technology Progress Award, respectively. He guided undergraduate students to win three national first prizes, five second prizes and three third prizes in the Innovation and Entrepreneurship Competition.

Siyuan Yang

Siyuan Yang is a Professor at the School of Resources and Environmental Engineering, Wuhan University of Technology, China. His research interests include mineral processing, micro-nano bubbles and secondary resource utilization. Prof. Siyuan Yang obtained his Ph.D. degree from the University of Queensland, Australia in 2018, and has worked in WUT since then. Prof. Siyuan Yang served as the director of the National Natural Science Foundation of China, National Key Research and Development Program of China, the Hubei Provincial Natural Science Foundation, the Open Fund of National Key Laboratory, the Open Fund of National Engineering Centre, and many other projects supported by enterprises. In the past five years, Prof. Siyuan Yang published more than 60 academic papers, including nearly 40 papers as the first author or corresponding author for international academic journals such as *Minerals Engineering, Chemical Engineering Science, and Applied Clay Science.*

He is a fellow of the China Nonferrous Metals Society, judging expert of the National Natural Science Foundation of China, youth Editorial Board Member of the *International Journal of Mining Science and Technology* and *Transactions of Nonferrous Metals Society of China*, and the Guest Editor of *Minerals*. He has granted more than 20 national invention patents, participated in writing 2 national standards, and received 5 provincial-level science and technology progress awards. Due to his outstanding contributions, Prof. Siyuan Yang has been elected as the Young Elite Scientists by the China Association for Science and Technology (CAST), and the Hubei Hundred Talent Program by Hubei Province, China.

Preface

Fine mineral flotation is a worldwide problem because fine mineral has a low mass, low probability of collision and adhesion between particles and bubbles, and is easy to entrain, making the flotation rate, recovery and concentrate grade low.

What is more, the high surface area of fine particles leads to the inevitable increase in ions in the solution. The mutual coagulation of fine minerals makes the slime covered and the high surface area increases the adsorbent dose of mineral particles, resulting in increased reagent loss and low separation efficiency.

Based on this, many scholars have carried out a large number of experimental studies, and have found that increasing the apparent diameter of ore particles and reducing the size of bubbles are two effective methods to solve the above-mentioned problems. This is because increasing the apparent diameter of ore particles can improve the collision probability between ore particles and bubbles, reduce the influence of mechanical entrainment and unavoidable ions, reduce the dosage of flotation reagents and improve the flotation index. When the bubble size is reduced, the gas-liquid interface area for particle adhesion increases and the mineralization efficiency increases.

Therefore, it is of great significance to increase the flotation efficiency of fine and low-grade minerals by bubble microfluidization and increasing the apparent diameter of particles.

We hope that "Advances in Fine Particles and Bubbles Flotation" will not only serve as a valuable resource for academia, but will also inspire curiosity and spark further inquiry into fine resource recovery theory and application.

Liuyi Ren, Wencheng Xia, Wei Xiao, and Siyuan Yang
Editors

minerals

Editorial

Editorial for Special Issue "Advances on Fine Particles and Bubbles Flotation"

Liuyi Ren

School of Resources and Environmental Engineering, Wuhan University of Technology, Wuhan 430070, China; rly1015@163.com or rly1015@whut.edu.cn

The froth-flotation process allows for the separation of solid particles according to differences in their physical and surface-chemistry properties. It is the most efficient and cost-effective separation process for particles within a narrow size range, especially that of minerals ranging from 10 to 100 μm. Since recent advances made in grinding, allowing for low-grade mineral deposits to be economically exploited, the flotation of fine particles (−10 μm) has become increasingly relevant. The recovery of fine particles by flotation is dependent on bubble–particle collisions, the probability of which decreases with both decreasing particle size and increasing the probability of detachment [1–5] and increases with increased particle size and decreasing bubble size. Currently, the most popular practice in froth floatation is particle collection via air bubbles [6,7], though it has been proven that nanobubbles can be used to improve the fine-particle flotation effect [8,9]. Many studies have also aimed to solve the fine-particle flotation problem by exploiting microbubble flotation equipment; for instance, the Reflux™ Flotation Cell was used to recover and clean fine hydrophobic particles, resulting in the stripping of fewer hydrophobic particles [10–13]. Further, it is well known that nanobubbles, referring to bubbles measuring only a few hundred nanometers wide, can extend the lower particle-size limits for the effective flotation of coal, phosphate, iron ore, and a number of other typical oxidized minerals [14–18]. With a wealth of research experience in microbubble-enhanced fine-particle flotation, the author is dedicated to demonstrating how nanobubbles offer an effective means of enhancing fine-particle flotation recovery, making it a subject of significant interest.

Thus, this Special Issue focuses on recent advances in fine particles and bubble flotation that are worth further study and application. Meanwhile, this collection provides guidance for those engaged in research on fine flotation, featuring studies including, but not limited to, the following topics: fine-particle flotation, microbubble flotation, nanobubble flotation, particle–bubble interaction, particle–bubble collision and adhesion, particle–bubble interface science, the aggregation of fine particle and bubbles, and the dynamic study of fine particles and bubble flotation systems. This Special Issue aims to contribute to the development of efficient and cost-effective fine-particle-enhanced flotation technologies for fine and ultrafine particles in the field of froth flotation.

One of the problems involved in fine-particle flotation is fine-gangue entrainment [9]. Mei et al. [19] studied the efficient separation of ultrafine coal using polyvinylpyrrolidone (PVP) as a regulator. They found that the addition of PVP improved the combustible recovery of clean coal and decreased the ash content, and that this effect was also presented by the selectivity index. Their study also revealed that the electrokinetic potential of minerals was sensitive to varying PVP concentrations, while the particle-size distribution and surface elemental compositions were also influenced by PVP. The findings of this research expand our understanding of the role of PVP as a regulator and provide a basis for the efficient separation of ultrafine coal.

The method of increasing mineral particle size is mostly used to solve problems associated with fine particles. Ming et al. [20] studied the sedimentation of fine arsenopyrite

Citation: Ren, L. Editorial for Special Issue "Advances on Fine Particles and Bubbles Flotation". *Minerals* **2024**, *14*, 439. https://doi.org/10.3390/min14050439

Received: 9 April 2024
Revised: 16 April 2024
Accepted: 17 April 2024
Published: 23 April 2024

with polyethyleneimine (PEI), a flocculant, and attempted to enhance fine-particle flotation by increasing its apparent size. They found that at approximately pH 7.5, after the addition of PEI-2, its recovery could be increased to 90%, while the maximum (D 100) and median (D 50) particle size of the arsenopyrite increased from 20 and 11 μm to 48 and 28 μm, respectively. The adsorption of PEI-2 on the arsenopyrite surface was caused by the chemisorption between the imino group and the active Fe/As sites. Applying PEI-2 to a fine disseminated arsenopyrite-type gold ore, a concentrate containing 36 g/t Au with a Au recovery of 88% could be obtained. Additionally, Guo et al. [21] studied the flocculation behavior of ultrafine silica particles in acid-leaching pulp using nonionic polymeric flocculants, finding that the tannic acid promotes the bridging flocculation of polyethylene oxide-quartz by forming associative complexes with larger clusters in solution, while montmorillonite electrostatically adsorbs on the quartz surface, enhancing its bridging with polyacrylamide. The low turbidity benefited from the higher shear resistance of the compact flocs structure, provided by polyethylene oxide/polyethylene oxide + tannic acid/polyacrylamide + montmorillonite. An efficient solid–liquid separation was achieved by using the synergistic flocculation of small molecule cofactors and polymer flocculants.

In the fine-particle separation process, the particle sizes of target and gangue minerals may be different, despite the lack of research on the flotation separation of two minerals with different particle sizes. Therefore, it is essential to further research fine-particle separations of gangue with different particle sizes. In Zhang et al.'s article [22], −10 μm rutile and −30 μm garnet particles were selected as samples. The authors proved that BHA and SPA changed the chemical environment of rutile but not garnet, and that SSF slightly affects the environment in which BHA and SPA interact with rutile. However, various reagents and reagent combinations have little effect on the surface-chemical environment of garnet, providing a certain basis for the flotation separation of fine rutile and garnet. In the future, in addition to flotation reagents, the separation of the two can also be studied in detail in terms of particle size. Otherwise, because of crystals' anisotropy, the exposed crystal-surface types of fine minerals are also important factors affecting fine-particle flotation. Moreover, as muscovite has a typical dioctahedral crystal structure, the atoms arranged in different directions of its crystal lattice cause the anisotropy of the physical and chemical properties of its crystal planes, leading to the anisotropy of these crystal planes in flotation. Ren et al. [23] studied the adsorption difference of octadecylamine on (002) and (131) crystal planes of fine muscovite, finding that (002) crystal planes have a higher surface energy and are more easily exposed than that of (131). Further, compared to Si–O bonds, Al–O bonds in muscovite had lower covalent-bond compositions and were easier to break, while O atoms acted as active sites in the flotation of muscovite, and the (131) crystal plane was more likely to adsorb with ODA than the (002) crystal plane. These results suggest that flotation efficiency can be improved by exposing more (131) crystal planes in the grinding stage.

In recent years, most studies in this area have been designed to determine the optimum conditions for the flotation of very fine minerals. For instance, Batjargal et al. [24] studied the "Frothing Performance of Frother-Collector Mixtures as Determined by Dynamic Foam Analyzer and Its Implications in Flotation", finding that the effects of frother type and concentration play a particularly significant role in optimizing the flotation conditions, alongside parameters such as particle size, morphology, and pH. Therefore, one of the most important issues to address for flotation is the effect of froth stability. Additionally, it has been found that bubble sizes become finer even at low concentrations of PPG600 and PPG400 frothers, while a significant decreases in bubble size have been found for collector–frother mixture systems, regardless of the concentration of the frother.

At present, scholars mainly study the relationship between nanobubbles and useful minerals, often ignoring the influence of bubbles on fine gangue minerals. In real-life applications, particles of the fine gangue mineral often enter foam products via bubbles, significantly affecting the quality of the concentrate. Lu et al. [25] studied the role of nanobubbles in the flotation behavior of hydrophilic serpentine. They found that the nanobubbles were stable, with no change in size and only a slight decrease in number as the resting time increased. When nanobubbles were introduced into a serpentine flotation, the presence of nanobubbles significantly reduced the flotation recovery of serpentine and reduced the froth entrainment rate of microfine-grained serpentine, which in turn reduced its flotation rate. In the depressant group trials, the nanobubbles also reduced the amount of depressant. In short, the presence of nanobubbles can prevent the floating of fine hydrophilic gangues during flotation.

Furthermore, process optimization, reagent development, and particle-size control are all important aspect for minimizing problems relating to fine-particles. Wang et al. [26] proposed using a scrubbing–desliming and flotation process to enrich vanadium, nickel, and molybdenum, and in a study obtained Vanadium–molybdenum-rich sludge and nickel-containing tailings products. The V_2O_5 and molybdenum grades in the sludge were 4.10% and 0.44%, respectively, and the recovery levels were 41.31% and 51.40%, respectively, while the nickel grade in the tailings was 1.49%. These products were roasted and leached. The vanadium, nickel, and molybdenum in the stone coal were effectively recovered through the beneficiation–metallurgy combination process, and the comprehensive utilization rate of the stone coal was improved. Wang et al.'s [27] study proved that sodium silicate (SS) can effectively improve the flotation-separation effect of bastnaesite and fluorite in salicylhydroxamic acid (SHA) systems. SS had a strong binding effect with the Ca site on the fluorite surface but a weak binding effect with the Ce site on the bastnaesite surface. In turn, SS can be used as an effective depressant in the flotation separation of fluorite and bastnaesite.

Mineral particle size is an important parameter in the mineral-beneficiation process. In industrial processes, the grinding process produces pulp with a qualified particle size for subsequent flotation processes. In this Special Issue, a hierarchical intelligent control method for mineral particle size based on machine learning is proposed by Zou et al. [28]. They present the method's practical application in the SAB-production process of an international mine to demonstrate its automation and intelligence, where the process throughput is increased by 6.05%, the power consumption is reduced by 7.25%, and the annual economic benefit is significantly improved.

Conflicts of Interest: The author declares no conflicts of interest.

References

1. Luttrell, G.H.; Yoon, R.H. Determination of the probability of bubble-particle adhesion using induction time measurements. In *Production and Processing of Fine Particles, Proceedings of the International Symposium on the Production and Processing of Fine Particles, Montreal, QC, Canada, 28–31 August 1988*; Pergamon: Oxford, UK, 1988; pp. 159–167.
2. Nguyen, A.V. The Collision between Fine Particles and Single Air Bubbles in Flotation. *J. Colloid Interface Sci.* **1994**, *162*, 123–128. [CrossRef]
3. Gontijo, C.D.F.; Fornasiero, D.; Ralston, J. The Limits of Fine and Coarse Particle Flotation. *Can. J. Chem. Eng.* **2007**, *85*, 562–785.
4. Miettinen, T.; Ralston, J.; Fornasiero, D. The limits of fine particle flotation. *Miner. Eng.* **2010**, *23*, 420–437. [CrossRef]
5. Farrokhpay, S.; Filippov, L.; Fornasiero, D. Flotation of Fine Particles: A Review. *Miner. Process. Extr. Metall. Rev.* **2021**, *42*, 473–483. [CrossRef]
6. Yoon, R.H. The role of hydrodynamic and surface forces in bubble-particle interaction. *Int. J. Miner. Process.* **2000**, *58*, 129–143. [CrossRef]
7. Nguyen, P.T.; Nguyen, A.V. Validation of the generalised Sutherland equation for bubble–particle encounter efficiency in flotation: Effect of particle density. *Miner. Eng.* **2009**, *22*, 176–181. [CrossRef]
8. Yoon, R.H. Microbubble flotation. *Miner. Eng.* **1993**, *6*, 619–630. [CrossRef]
9. George, P.; Nguyen, A.V.; Jameson, G.J. Assessment of true flotation and entrainment in the flotation of submicron particles by fine bubbles. *Miner. Eng.* **2004**, *17*, 847–853. [CrossRef]
10. Cole, M.J.; Dickinson, J.E.; Galvin, K.P. Recovery and cleaning of fine hydrophobic particles using the Reflux™ Flotation Cell. *Sep. Purif. Technol.* **2020**, *240*, 116641. [CrossRef]
11. Galvin, K.P.; Dickinson, J.E. Fluidized bed desliming in fine particle flotation—Part II: Flotation of a model feed. *Chem. Eng. Sci.* **2014**, *108*, 299–309. [CrossRef]

12. Galvin, K.P.; Harvey, N.G.; Dickinson, J.E. Fluidized bed desliming in fine particle flotation—Part III flotation of difficult to clean coal. *Miner. Eng.* **2014**, *66–68*, 94–101. [CrossRef]
13. Dickinson, J.E.; Galvin, K.P. Fluidized bed desliming in fine particle flotation—Part I. *Chem. Eng. Sci.* **2014**, *108*, 283–298. [CrossRef]
14. Ren, L.; Zhang, Z.; Zeng, W.; Zhang, Y. Adhesion between nanobubbles and fine cassiterite particles. *Int. J. Min. Sci. Technol.* **2023**, *33*, 503–509. [CrossRef]
15. Fan, C.; Ren, L.; Zhang, Y.; Bao, S. Grinding effect of sodium silicate on muscovite and its mechanism analysis. *Miner. Eng.* **2023**, *199*, 108106. [CrossRef]
16. Zhang, Z.; Ren, L.; Zhang, Y. Role of nanobubbles in the flotation of fine rutile particles. *Miner. Eng.* **2021**, *172*, 107140. [CrossRef]
17. Wei, P.; Ren, L.; Zhang, Y.; Bao, S. Influence of Microbubble on Fine Wolframite Flotation. *Minerals* **2021**, *11*, 1079. [CrossRef]
18. Zhu, S.; Ren, L.; Bao, S.; Zhang, Y.; Chen, B. Effect of different particle size fractions on flotation separation of fine rutile from garnet. *Adv. Powder Technol.* **2024**, *35*, 104396. [CrossRef]
19. Mei, Y.; Lin, Q.; Wu, C.; Huang, W.; Cao, D.; Liu, K. Efficient Separation of Ultrafine Coal Assisted by Selective Adsorption of Polyvinylpyrrolidone. *Minerals* **2022**, *12*, 725. [CrossRef]
20. Ming, P.; Zou, D.; Li, F.; Xing, Q.; Wang, Z. Sedimentation of Fine Arsenopyrite with PEI and the Flotation Significance. *Minerals* **2022**, *12*, 824. [CrossRef]
21. Guo, B.; Zhan, X.; Jiang, K.; Xie, H.; Deng, R. Flocculation Behavior of Ultrafine Silica Particles in Acid Leaching Pulp by Nonionic Polymeric Flocculants. *Minerals* **2023**, *13*, 582. [CrossRef]
22. Zhang, Z.; Ren, L.; Zhang, Y.; Bao, S. Microflotation of Fine Rutile and Garnet with Different Particle Size Fractions. *Minerals* **2022**, *12*, 1238. [CrossRef]
23. Ren, L.; Jiao, Z.; Zhang, Y.; Bao, S. Adsorption Difference of Octadecylamine on (002) and (131) Crystal Planes of Fine Muscovite and Its Guidance on Fine Muscovite Flotation. *Minerals* **2023**, *13*, 519. [CrossRef]
24. Batjargal, K.; Guven, O.; Ozdemir, O.; Karakashev, S.; Grozev, N.; Boylu, F.; Çelik, M. Frothing Performance of Frother-Collector Mixtures as Determined by Dynamic Foam Analyzer and Its Implications in Flotation. *Minerals* **2023**, *13*, 242. [CrossRef]
25. Lu, B.; Xu, W.; Luo, C.; Li, W.; Su, X.; Song, Y.; Zhou, J.; Li, K. Effect of Nanobubbles on the Flotation Behavior of Microfine-Grained Serpentine. *Minerals* **2023**, *13*, 1299. [CrossRef]
26. Wang, M.; Cai, L.; Wen, J.; Li, W.; Yang, X.; Yang, H. The Prospect of Recovering Vanadium, Nickel, and Molybdenum from Stone Coal by Using Combined Beneficiation and Metallurgy Technology Based on Mineralogy Features. *Minerals* **2023**, *13*, 21. [CrossRef]
27. Wang, M.; Xiong, W.; Xiao, J.; Guo, Y.; Deng, J.; Chen, D.; Ouyang, A.; Lei, M.; Zhang, L. Selective Adsorption of Sodium Silicate on the Surface of Bastnaesite and Fluorite in Salicylhydroxamic Acid System under Alkaline Conditions. *Minerals* **2023**, *13*, 69. [CrossRef]
28. Zou, G.; Zhou, J.; Song, T.; Yang, J.; Li, K. Hierarchical Intelligent Control Method for Mineral Particle Size Based on Machine Learning. *Minerals* **2023**, *13*, 1143. [CrossRef]

minerals

MDPI

Article

Efficient Separation of Ultrafine Coal Assisted by Selective Adsorption of Polyvinylpyrrolidone

Yujie Mei [1,2], Qiuyu Lin [1,2], Changning Wu [2,3,4], Wei Huang [2,3,4], Daofan Cao [5,6] and Ke Liu [2,3,4,*]

1 School of Chemistry and Chemical Engineering, Harbin Institute of Technology, Harbin 150001, China; 11949014@mail.sustech.edu.cn (Y.M.); 11849610@mail.sustech.edu.cn (Q.L.)
2 Department of Chemistry, Southern University of Science and Technology, Shenzhen 518055, China; wucn@sustech.edu.cn (C.W.); huangw8@sustech.edu.cn (W.H.)
3 Shenzhen Engineering Research Center for Coal Comprehensive Utilization, School of Innovation and Entrepreneurship, Southern University of Science and Technology, Shenzhen 518055, China
4 Clean Energy Institute, Academy for Advanced Interdisciplinary Studies, Southern University of Science and Technology, Shenzhen 518055, China
5 Birmingham Centre for Energy Storage (BCES), University of Birmingham, Birmingham B15 2TT, UK; dxc816@student.bham.ac.uk
6 School of Chemical Engineering, University of Birmingham, Birmingham B15 2TT, UK
* Correspondence: liuk@sustech.edu.cn; Tel.: +86-0755-8801-0097

Abstract: The efficient separation of ultrafine coal is a challenging process due to the ultrafine particle size and gangue entrainment. In this study, a polymer, polyvinylpyrrolidone (PVP), was introduced as a regulator for ultrafine coal beneficiation. The addition of PVP improved the combustible recovery of clean coal and decreased the ash content. This effect was also presented by the selectivity index. The regulation mechanism of PVP was investigated using diverse methods. The adsorption tests performed demonstrated the adsorption amount of PVP on coal, kaolinite, and quartz, which were related to the increase in the separation efficiency. A zeta potential analyzer was employed to elucidate the effect of PVP on the electrical properties of ultrafine particles. The results revealed that the electrokinetic potential of mineral was sensitive to the varying PVP concentration. The particle size distribution was observed to value the influence of PVP on the particle behavior, which was tested by a laser particle size analyzer. X-ray photoelectron spectroscopy was used to investigate the surface elemental compositions of coal, kaolinite, and quartz, which were regulated by the adsorption of PVP. This research is beneficial to understanding the role of PVP as regulators and provides a basis for the efficient separation of ultrafine coal.

Keywords: polyvinylpyrrolidone; ultrafine coal; adsorption; separation efficiency

Citation: Mei, Y.; Lin, Q.; Wu, C.; Huang, W.; Cao, D.; Liu, K. Efficient Separation of Ultrafine Coal Assisted by Selective Adsorption of Polyvinylpyrrolidone. *Minerals* **2022**, *12*, 725. https://doi.org/10.3390/min12060725

Academic Editors: Liuyi Ren, Wencheng Xia, Wei Xiao, Siyuan Yang and Kevin Galvin

Received: 15 April 2022
Accepted: 2 June 2022
Published: 6 June 2022

1. Introduction

As the most abundant fossil fuel in the world, coal accounts for over 50% of China's total energy consumption. Coal washing, which results in a large amount of slime, has put the spotlight on the utilization of coal cleaning technologies [1]. However, coal slime is difficult to recycle, and the majority of coal slime produced is classified as solid waste. Thus, the aim of coal slime recycling is not only save energy but also protect the environment [2]. Coal slime generated by coal washing plants usually has a high content of gangue minerals, with a fine or ultrafine particle size [3]. Fine coal are particles with particle sizes less than 0.5 mm while particles with diameters less than 0.1 mm are defined as ultrafine [4].

Froth flotation is a relatively effective method used to separate fine particles [5] and is widely used for fine coal separation in many countries, including China. However, the reprocessing of ultrafine coal by flotation is often not economically viable. The grain size is an important factor in flotation, especially for ultrafine particles, which need a longer processing time, higher energy, and more reagent consumption, to obtain a good separation performance [6]. Since hydrophobic coal particles tend to float on the water surface, partial gangue particles can be embedded in coal, causing gangue entrainment. Meanwhile, the gap between coal and gangue narrows down at ultrafine grain sizes [7]. Attempts have thus been made to improve ultrafine coal flotation. Methods to enhance flotation can be divided into physical and physicochemical methods, such as changing the flow pattern of the air supply [4]; pre-conditioning coal slurry with microbubbles [8], nanobubbles [9], and high-voltage treatment [10]; carrier flotation [11]; selective oil agglomeration [12]; and the introduction of a mixed reagent system [13].

For the secondary recycling of carbonaceous resources from coal slime, the introduction of a flotation reagent has been considered as a research hotspot and an effective antidote to improve the separation efficiency. Traditional hydrocarbon oil collectors, such as diesel oil and kerosene, are widely used for coal flotation. However, oil consumption during coal slime separation is very high [14]. The adhesion of reagents on particles is a complicated process and is affected by the grain size, the type of additives used, and the pH of the solution [15,16]. It is difficult to obtain good adhesion or adsorption of non-polar oil collectors on ultrafine coal, wherein a limited efficiency is achieved due to the ultrafine grain size and particle–bubble attachment [17]. Due to the application of large machinery and the rapid dissociation rate of coal particles during modern coal processing, the high surface roughness and high ash content deteriorate the performance of ultrafine coal flotation [18]. Some surfactants have been proposed to improve the performance of coal recycling. Xia et al. used a cationic surfactant called dodecyl trimethyl ammonium bromide to enhance the adsorption of oil collectors on the coal surface, which improved the collection efficiency of dodecane limited by the oxygen functional groups on the coal surface [19]. Temperature also plays a key role in the adsorption of surfactants: Marsalek et al. found that the adsorption of cetyltrimethylammonium bromide on coal was influenced by temperature [20]. The introduction of surfactants plays a similar role in coal surface modification, adjusting the proportion of coal surface functional groups and reducing the effect of the coal surface roughness on the adhesion process [21]. It can be concluded that compound reagents contribute to an improved regulation performance in coal flotation. Numerous studies have demonstrated the superiority of compound agents, whereas polymer reagents have often been ignored. For fine particles, polymer reagents possess the advantages of a lower dosage and better adhesion efficiency. Polymer reagents are currently attracting increasing attention. Zou et al. reported that the introduction of a polymer, such as hydrophobically modified polyacrylamide, improved the combustible recovery of coal flotation [22]. Li et al. took advantage of the collaboration between nanobubbles and polyaluminum chloride to efficiently separate fine coal, with the nanobubbles promoting the flotation recovery while polyaluminum chloride inhibited entrainment by increasing the grain size [23]. Lv et al. synthesized a novel pH-sensitive flocculant and studied its effect on fine particles [24]. Studies exploring the impact of polymer reagents on fine coal separation are burgeoning. However, it is crucial to improve the floatability and selectivity of ultrafine coal by developing novel reagents.

As a novel polymer reagent rarely used in coal separation, polyvinylpyrrolidone (PVP) was adopted in this study. PVP has been adopted in the pharmaceutical and petroleum industries as an additive for separation, and acts as a carrier of synthetic materials. Palchoudhury et al. used PVP nanoparticles as a carrier to separate oil from wastewater and recycled petroleum wastewater [25]. Alabresm et al. investigated the synergistic effect of PVP nanoparticles and oil-degrading bacteria for the remediation of oil-water mixtures [26]. To date, only a few studies have explored the impact of PVP on ultrafine coal separation. PVP plays a key role in many separation scenarios. Through an experimental investigation, this study proves that PVP can facilitate the separation of ultrafine coal.

To understand how PVP facilitates ultrafine coal separation, laboratory flotation tests were conducted to observe the influence of PVP on the separation performance. A selectivity index (*SI*) was introduced to measure the results. Subsequently, an ultraviolet spectrophotometer (Uv-vis) was used for the adsorption test, wherein the purified minerals were used to test the adsorption selectivity of PVP on coal, kaolinite, and quartz. The agglomeration behavior of the ultrafine particles with the introduction of PVP was investigated using a laser particle size analyzer (LPSA). The electrokinetic potential of the particles was measured using a zeta potential analyzer, and X-ray photoelectron spectroscopy (XPS) was employed to analyze the surface energy change of the particles influenced by PVP. We hope that the results of this study will prove beneficial in terms of understanding the role polymer reagents play in ultrafine coal separation.

2. Materials and Methods

2.1. Materials

The coal slime (ultrafine coal) and clean coal samples were procured from RecoTech Co., Ltd. (Tsingtao, China). Kaolinite and quartz were purchased from Beijing Mairuida Technology Co., Ltd. (Beijing, China). To understand the basic details of coal slime, we carried out proximate and elemental analyses (Table 1). We observed that the ash content of coal slime was 55%, and the ash content of clean coal was lower than 8%. The gross calorific value of clean coal was 24.02 MJ/kg, and its vitrinite reflectance was less than 0.5%. It could thus be classified as sub-bituminous coal according to ISO 11760. To identify the primary gangue mineral composition in coal slime, the ultrafine coal was scanned by X-ray diffraction (XRD, Rigaku Smart Lab, Tokyo, Japan). The results showed that the main gangue components were quartz and kaolinite (Figure 1). An LPSA (Beckman, LS 13320, Brea, CA, USA) was used for the particle size distribution analyses (Figure 2). The results showed that the mean size (D_{50}) of coal slime was 23.73 μm, with the size of 90% of the fine particles (D_{90}) being 76.91 μm, which was much lower than 0.1 mm. Thus, the coal slime was ultrafine coal. To further explore the adsorption and agglomeration properties of the ultrafine particles, clean coal, kaolinite, and quartz were also ground to the ultrafine grain size. Correspondingly, the size distribution of clean coal, quartz, and kaolinite is shown in Figure 3. The D_{50} of clean coal, quartz, and kaolinite is 24.11, 6.52, and 5.51 μm, respectively.

Table 1. Proximate and elemental analyses of coal slime (ultrafine coal).

	Industrial Analysis/%				Elemental Analysis/%				
chemical	M_{ad}	A_{ad}	V_{daf}	FC_{daf}	C_{daf}	H_{daf}	N_{daf}	S_{daf}	O_{daf} *
results	2.03	55.0	23.41	64.58	65.19	5.34	1.04	1.61	26.82

Note: M—moisture; A—ash; V—volatile; FC—fixed carbon; ad—air-dry basis; daf—air-dry ash free basis; * by difference.

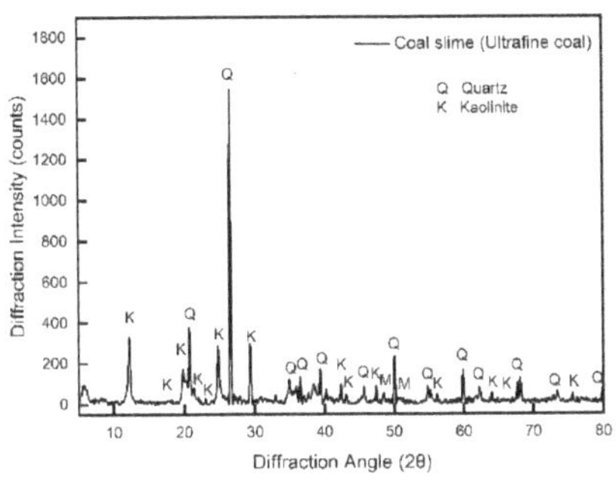

Figure 1. XRD result of coal slime (ultrafine coal).

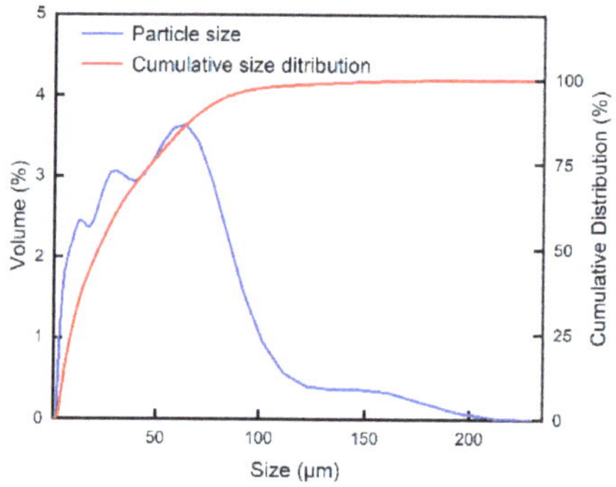

Figure 2. Size distribution of coal slime (ultrafine coal).

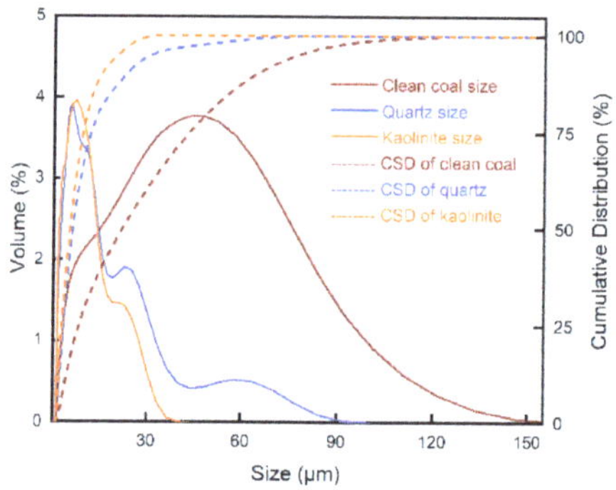

Figure 3. Size distribution of clean coal, quartz, and kaolinite.

All the reagents used in the experiments (purchased from Shanghai Macklin Biochemical Co., Ltd., Shanghai, China) are listed in Table 2. The solvent used in this study was deionized water (Milli-Q Integral 5, France), with a resistivity of less than 18.2 MΩ cm.

Table 2. List of reagents.

Name	Purity	Molecule Weight	Abbreviation
dodecane	≥99%	170	DD
polyvinylpyrrolidone		27,000	PVP
2-octanol	≥99%	130	2-octanol

Polyvinylpyrrolidone $(C_6H_9NO)_n$ is a non-ionic polymer The average molecule weight of PVP used in this paper was 27,000. Its partial macromolecular structure is shown in Figure 4.

2.2. Laboratory Flotation Test

The laboratory flotation tests of ultrafine coal were carried out in an XFD flotation cell purchased from Wuhan Exploration Machinery Co., Ltd. (Wuhan, China). The flotation cell volume was 1 L. In total, 80 g of coal slime and 800 mL of water were added to prepare the suspension (8 wt%). Dodecane (DD) was used as the collector, 2-octanol was used as the frother, and PVP (used as the regulator) was dispersed into the slurry before flotation. The PVP solution was introduced at several concentrations (1, 3, 6, 10, 15 mg/L) to test its effect on recovery. PVP was added for 1 min of preset mixing, then DD (0.064 mg/mL) and 2-octanol (0.04 mg/mL) were added to the flotation cell. For all the flotation tests, the stirring speed of the agitator was maintained at 1500 rpm and the air supply rate was set to 0.15 m^3/h. The flotation cell was filled to the level of the standard waterline with tap water (pH = 6.8). All the tests were conducted at 25 °C. The separation products were collected, filtered, and dried in a ventilated drying oven at 95 °C. Proximate analysis was then conducted to clear the ash content of the samples.

The recovery of flotation was calculated using Equation (1):

$$R = \frac{M_f}{M_t} \cdot 100\% \tag{1}$$

where M_f is the dried weight of combustible matter (mainly coal) or ash (mainly quartz and kaolinite), M_t is the weight of the total feedstock of flotation cell (weight of coal slime), and R is the flotation recovery.

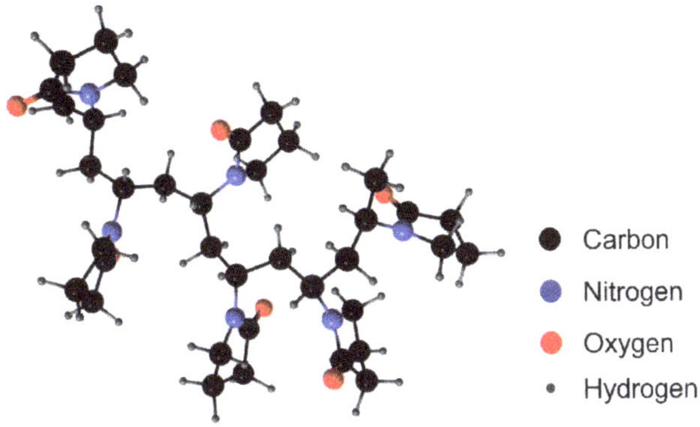

Figure 4. Partial macromolecular structure of PVP [27].

The flotation recovery can reflect the yield of clean coal. However, some further explanation is required for the separation selectivity. Thus, we first calculated the flotation rate. Some scholars have assumed that the flotation process can be described as a first-order reaction [28]. The first-order dynamics equation assumes that all matter floats at the same rate. In fact, due to differences in the grain size and mineral species, particles should have different flotation parameters. This study used Kelsall's model to calculate the flotation rate as shown in Equation (2) [27]:

$$R = R_{max} - [R_{fast} \cdot e^{-k_f t} + \left(R_{max} - R_{fast}\right)] \cdot e^{-k_s t} \tag{2}$$

where R_{max} is the maximum theoretical recovery, R_{fast} is the maximum recovery of the fast floating fraction, and k_f and k_s are the rate constant of the fast float and slow float components, respectively. For practical purposes, flotation separation is often carried out using multi-stage circulation. The fast-float stage often plays the reference role for the particle cycling duration time in the flotation circuit while the slow stage is not worth saving. Thus, we mainly analyzed the parameters of the fast stage, which was more reasonable. Therefore, the rapid stage is usually more important, where R_{max} and k_f determine the efficiency. The solver tool in Excel was used for mathematical analysis of the fitting process, and the constraints for the parameters were added ($R_{fast} \leq 100, R_{fast} > R_{slow}, k_f > k_s$). The rate function was set such that the operation represented the true physical process with the fast and slow stage. The modified rate constant (K_m, min^{-1}) was introduced to value the response of different components as shown in Equation (3) [29]:

$$K_m = \frac{R_{max} \cdot k_f}{100} \tag{3}$$

We used the SSQ function to fit the results. The formula is Equation (4):

$$SSQ = \sum (R - \check{R})^2 \tag{4}$$

where R is the real recovery and $\check{R}$ is the fitted value.

Finally, we introduce the selectivity index (SI) to compare the flotation selectivity of serial tests. The SI was calculated using Equation (5) [30]:

$$SI\left(\frac{combustible}{ash}\right) = \frac{K_{m,com}}{K_{m,ash}} \tag{5}$$

where $K_{m,com}$ and $K_{m,ash}$ represent the modified rate constants of the combustible matter (mainly coal) and ash materials, respectively.

2.3. Adsorption Amount and Equilibrium Test

This study employed Uv-vis (UV-2600, Shimadzu, Japan) to test the adsorption characteristics of PVP. Before conducting the test, PVP was prepared as solutions with different concentrations (1, 3, 6, 10, 15, and 30 mg/L). Subsequently, 4 g of clean coal, quartz, and kaolinite were dispersed in 50 mL of PVP solution (8 wt%). The PVP solutions with a concentration of 6 mg/L were used to analyze the rate of adsorption. Pure PVP solutions without the solid samples were first scanned from 150–500 nm using Uv-vis to achieve the maximum absorbance value. The value of the peaks was consistently located at 194 nm. At the same time, the absorbance of PVP solutions with different concentrations was recorded to draw the calibration curve. Thereafter, the clean coal, kaolinite, and quartz particles were added to the PVP solutions. These mixtures were stirred at 500 rpm for 3 h to ensure sufficient adsorption of PVP on particles, which were then transferred into the centrifuge tube for a 10 min centrifugation at 10,000 rpm by High-Speed Centrifuge (Sorvall LYNX 6000, Thermo Fisher Scientific, Waltham, MA, USA). Then, the supernates were then extracted using a syringe and used for the Uv-vis test. The residual amount of PVP in the supernates was calibrated using the calibration curve, and the adsorption quality of PVP on the particles was deduced by subtraction. The adsorption amount of PVP on the particles was calculated by Equation (6):

$$A = [(c_o - c_r + c_b) \cdot V \cdot 1000]/m \tag{6}$$

where A (mg/g) is the adsorption mass of PVP on the particles, c_o (mg/L) is the original concentration of the PVP solution used for the adsorption test, c_r (mg/L) is the residual concentration of PVP in the supernate after centrifugation, c_b is the dissolved matter of the solid particles in the supernate, V (mL) is the volume of the PVP solution, and m (g) is the corresponding net weight of the particles in the supernate. This method is called the subtraction method [31].

2.4. Zeta Potential Test

A zeta potential analyzer (Malvern Instruments, Malvern, Britain) was used to evaluate the properties of the ultrafine particles in terms of the surface charge and zeta potential value. Deionized water was used as a solvent, and 0.01 mol/L of KCL solution was prepared as the background electrolyte [32]. The pH during zeta potential measurement in this study was maintained at 6.8. PVP was completely dissolved in water and no flocs were observed. To understand the effects of the PVP concentration on coal, quartz, and kaolinite, we prepared eight different concentrations of the PVP solution (1, 2, 3, 6, 10, 15, 20, and 30 mg/L). Considering the increase in the self-aggregation and decrease in the water solubility of PVP with higher concentrations, concentrations greater than 30 mg/L were excluded from the test range. On the other hand, this study focused on the effects of the selective adsorption of PVP on ultrafine particles. Excessive PVP may induce supersaturated adsorption, thus reducing its adsorption selectivity and changing the role of

PVP in ultrafine coal separation. In total, 4 g of the solid sample (clean coal, kaolinite, and quartz) was added to 50 mL of the PVP solution and stirred for 3 h at 25 °C. Subsequently, 5 mL of the mixture was transferred into the centrifuge tube and centrifuged for 10 min at 10,000 rpm using the high-speed centrifuge (Thermo Fisher scientific, Sorvall LYNX 6000, Waltham, MA, USA). Thereafter, the supernate in the centrifuge tube was filtered and the filtrate was used for analysis. For each filtrate sample, tests were conducted several times with n repetition, and the standard error(y) of the zeta potential value was calculated by the STDEV function, given in Equation (7):

$$y = \sqrt{\frac{\sum(x - \overline{x})}{(n-1)}} \times 100\% \tag{7}$$

where x is the test value and ($\overline{x}$) is the average value for one point, and n is the number of test repetitions for one point. The value of y was set as the depth of the error bars and used to analyze the validity of the data.

2.5. Particle Size Distribution Test

The LPSA was used to evaluate the flocculation behavior of the ultrafine particles. The test system worked well in the wet method. Regarding the optical parameter of the analysis system, the real part of the medium refractive index was set to 1.8, 1.6, and 1.5 for coal, kaolinite, and quartz, respectively. Deionized water was used for sample dispersion and circulation, and an 8 wt% ultrafine particle suspension with 6 mg/L PVP was stirred at a rotation speed of 500 r/min for 3 h. Afterward, the dispersed sample was injected into the test system of the LPSA with a syringe. Agitation and ultrasonic defoaming were continued during the measurement [27].

2.6. XPS Test

XPS was used for the elemental analysis of clean coal, kaolinite, and quartz. To prepare the test samples with the PVP treatment, 4 g of clean coal, kaolinite, and quartz were added to 50 mL of PVP solution and stirred for 3 h at 25 °C. Then, 5 mL of the mixture was centrifuged and dried. To detect the surface changes, XPS was performed at 25 °C (ULVAC PHI 5000 VersaProbe III, Chigasaki, Japan). From a monochromatic X-ray source, the Al Kα radiation (hν = 1486.6 eV) and a light spot size of 200 μm were set [33]. The spectra were recorded and stored at the pass energy of 26 eV in steps of 0.05 eV.

3. Results and Discussion

3.1. Laboratory Flotation

The mass recovery and flotation rate are used to evaluate flotation performance. We set up a series of flotation tests with different PVP concentrations and collected their separation products to calculate the recovery and flotation rate parameters. The results (Table 3 and Figure 5) revealed a difference in the flotation performance induced by PVP. The recovery and kinetic parameters (R_{max}, k_f, and K_m) were calculated and fitted by Equations (1)–(5). As seen in Table 3, the results indicated an increase in the combustible recovery of froth products with the introduction of PVP. Low dosages of PVP can relieve the constraints for ultrafine coal flotation recovery and can prove beneficial for reducing the number of oil collectors and energy consumption. The increase in the flotation rate proved that PVP facilitated the flotation of ultrafine particles, which can be attributed to the regulating role of PVP in particle behavior [27]. Meanwhile, the change in the ash recovery in the froth products was more complicated. In the tests with 1, 3, and 6 mg/L of PVP, the ash content appeared to be lower than the value in the test without PVP (blank control group). This seems to suggest that a low dosage of PVP is more likely to increase the flotation of combustible matter compared to the ash component. However, the maximum ash recovery of the flotation test with 20 mg/L PVP tended to approach the results of the blank control group. This difference could be attributed to the adsorption characteristics of PVP. Excessive

PVP can lead to the heterogeneous coagulation of coal and gangue particles. The modified rate constant K_m takes both the maximum recovery and flotation rate into consideration, which is considered fair regarding process evaluation. The value of $K_{m,com}$ increased with a higher PVP concentration; on the contrary, $K_{m,ash}$ was reduced with a higher PVP addition. The increase in the flotation rate constant can be attributed to the enhancement in the collision probability between the particles and bubbles [29]. The role of the reagents in this process was mainly focused on the surface modification of particles [34]. The agglomeration and adsorption behavior also provide evidence for the regulating role of PVP later in this study. Notably, when PVP exceeded a reasonable amount, the increase in the combustible recovery and ash recovery occurred simultaneously. More ash components disseminated in the froth products, limiting the further increase in $R_{max,com}$ and $K_{m,com}$. To describe the changes and measure the impact of PVP on the separation efficiency, we calculated the *SI* (Equation (5)) of the flotation tests, and plotted it against other flotation indexes.

Table 3. Flotation kinetic parameters (R_{max}, k_f, and K_m) with varying PVP concentrations.

PVP (mg/L)	Combustible Matter			Ash		
	R_{max}	k_f	$K_{m,com}$	R_{max}	k_f	$K_{m,ash}$
0	41.36	3.13	1.29	28.37	0.14	0.04
1	43.93	3.28	1.44	23.88	0.11	0.03
3	59.77	4.16	2.49	21.51	0.08	0.02
6	84.58	3.23	2.73	22.08	0.08	0.02
10	78.76	3.67	2.89	28.54	0.12	0.03
20	80.01	3.62	2.90	35.69	0.10	0.04

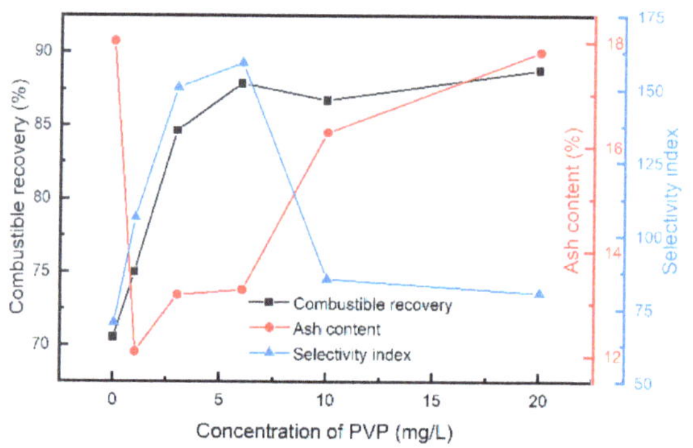

Figure 5. Combustible recovery (gray), ash content (red), and selectivity index (blue) of coal slime flotation with various concentrations of PVP.

Figure 5 shows the results of the combustible recovery and ash content of the froth products, and the calculated *SI* of various tests. As a blank control group, the tests without PVP yielded the highest ash content (18.02%) in the froth products, and the lowest combustible recovery (38.71%) and *SI* (70.22). Compared to the blank control group, the lowest ash content (12.08%) was achieved with 1 mg/L PVP. However, the combustible recovery and *SI* did not yield perfect results (70.53% and 106.21, respectively). The best *SI* (159.03) was obtained in the test conducted with 6 mg/L PVP; the ash content (13.17%) and combustible recovery (87.89%) were also great. The best maximum combustible recovery (88.81%) was obtained by the test performed with 20 mg/L PVP; however, the *SI* (85.26) was poor due to the increased entrainment of ash content (16.27%). Both the selectivity and recovery need to be considered when improving efficiency. Relatively good results were achieved with the 1 and 3 mg/L PVP concentrations. In general, particles tend to be coated by reagents, and the radius of particles and the length of the reagent play important roles in flotation [35]. As a type of polymer, PVP has an advantage regarding the length of its molecular chain, which is effective in enhancing ultrafine particle flotation.

3.2. Adsorption Amount and Equilibrium

Uv-vis was used to test the adsorption capacity of PVP on the clean coal, kaolinite, and quartz particles. A hydration film typically exists on the surface of particles to prevent the collision or adhesion of particles, bubbles, or reagents. Effective adhesion is achieved only if the film is ruptured. Once stable adsorption is achieved, the amount of adsorption reflects the adsorption capacity. The results in Figure 6 prove that PVP adsorbed a less amount on coal than quartz and kaolinite regardless of the scale of the initial concentration of PVP (Figure 6a) or the equilibrium time (Figure 6b). In addition, the saturated adsorption amount of PVP on the particle surface, which is a critical parameter for determining the adsorption property, was closely related to the specific surface area of the particle. Although the saturated adsorption amount of PVP on kaolinite and quartz was greater than that on clean coal, the specific surface areas of kaolinite and quartz were much higher than on coal. It is worth mentioning that when the time increased, the adsorption of kaolinite and quartz increased correspondingly. At this moment, the initial concentration of PVP (6 mg/L) was more beneficial to the adsorption of PVP on coal. One explanation is that the adsorption strength of PVP could be stronger than that of gangue minerals, and the saturation adsorption amount was limited by the specific surface area of coal. However, the contact of the reagents with the particles in flotation was more complicated, wherein the adsorption efficiency was also influenced by the hydromechanical characteristics. Thus, the saturation adsorption of kaolinite and quartz was hardly achieved in the suspension adjustment process during flotation due to the process duration time. This could be the main reason for the improvement in the combustible recovery for low PVP dosages, and the increase in gangue entrainment can be attributed to the increase in PVP adsorption for high PVP dosages. Finally, it can be deduced that the enhancement in flotation was mainly related to the adsorption of PVP on the particles.

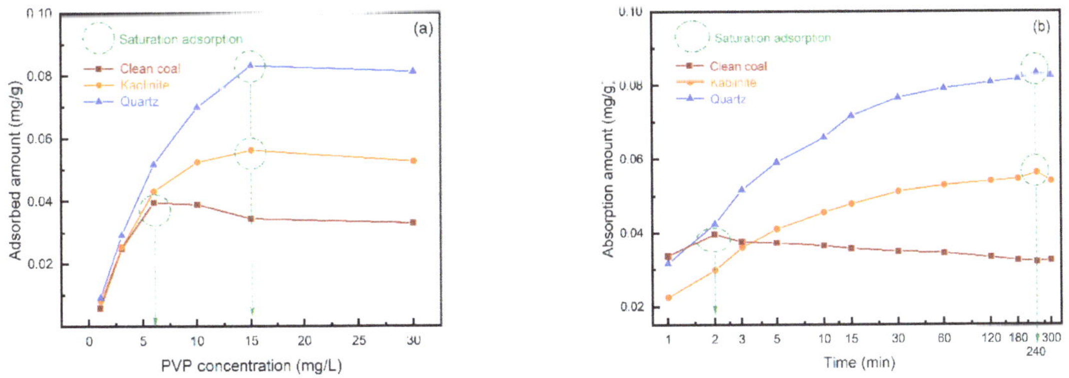

Figure 6. Adsorption amount of PVP on clean coal (wine), kaolinite (orange), and quartz (blue): (**a**) adsorption amount with an increase in the PVP concentration amd (**b**) adsorption amount with an increase in time.

3.3. Zeta Potential

The measured zeta potential values of the clean coal, kaolinite, and quartz suspensions at different PVP concentrations are shown in Figure 7, where the error bars represent the standard deviation of the three replicate tests. As a widely used metric for analyzing the interaction of particles or macromolecules in the field of surface chemistry, the zeta potential explains changes in the electrostatic attraction or repulsion induced by PVP [36]. With 0.01 mol/L KCL, the mean value of the initial zeta potentials of kaolinite and quartz were

−37.68 and −42.41 mV, respectively, both of which were much higher than the −28.88 mV of clean coal. With an increase in the dosage of PVP, the zeta potential of clean coal fluctuated from −26.95 to −28.88 mV. The response was much weaker than the change noticed in gangue minerals. The zeta potential of kaolinite increased to −27.41 mV, whereas the zeta potential of quartz increased to −31.25 mV and then remained constant at the PVP concentration of 30 mg/L, which caused a strong electrostatic attraction between the gangue minerals. Kaolinite and quartz tended to agglomerate from the analysis of the homogeneous system. However, the zeta potential was measured under relatively ideal conditions of pure matter. In a real system of ultrafine coal separation, competitive adsorption of PVP can occur between different particles, and particle behavior is associated with the dosage of reagents. The ideal separation effect will only be maintained when the right range of reagent concentrations is used.

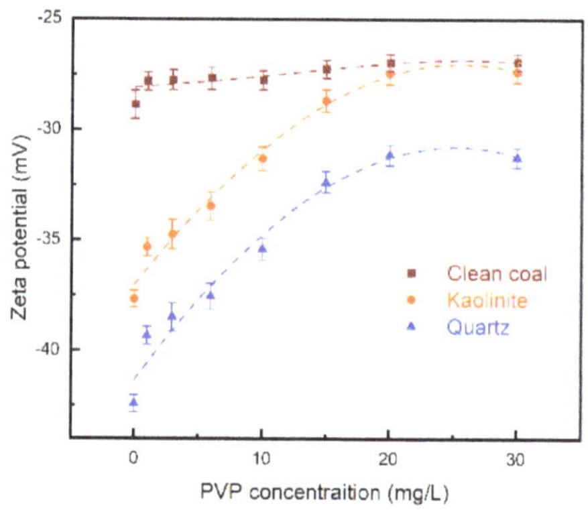

Figure 7. Zeta potentials of clean coal (wine), kaolinite (orange), and quartz (blue) with different PVP concentrations.

3.4. Particle Size Distribution

To explain the effect of PVP on the particle size distribution of clean coal, kaolinite, and quartz, we chose the PVP solution of 6 mg/L for the LPSA measurement, corresponding to the group that yielded the best *SI* results in the flotation test. The LPSA functions through light diffraction and scattering of particles. After eliminating the effect of the test environment and sample characteristics, an agglomeration tendency was noticed in all three types of particles with the assistance of PVP, as shown in Figure 8. The median size (D_{50}) was often used to describe the feature size of a particle swarm. On the one hand, the D_{50} of the clean coal particles increased from 24.11 to 36.56 μm after PVP adsorption, which was beneficial for combustible recovery. On the other hand, the D_{50} of the kaolinite and quartz particles was below 10 μm (Figure 9). In comparison, the coal flocs tended to float upward after the adsorption of PVP, and the entrainment of kaolinite and quartz was depressed. This condition was consistent with the flotation behavior of ultrafine coal with the 6 mg/L PVP addition. Apart from D_{50}, D_{90} represents the maximum diameters of 90% particles, which also presented growth. All the changes observed in the particle size distribution prove the effect of PVP adsorption on the particle behavior. By observing the adsorption amount of PVP on the surface of clean coal, kaolinite, and quartz, we formed the opinion that the addition of a low PVP dosage improves the selectivity during the flotation of ultrafine coal.

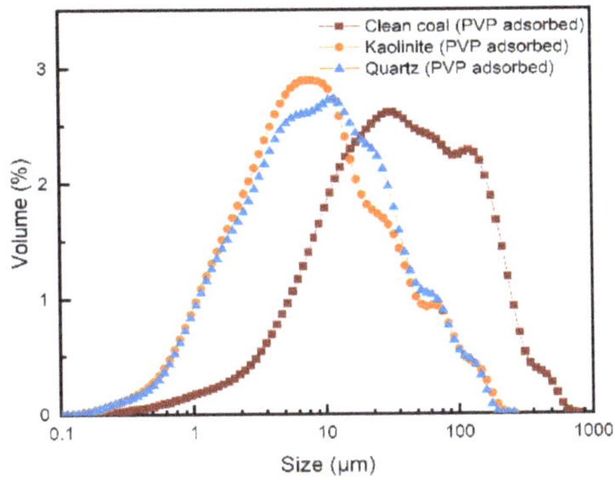

Figure 8. Size distribution of clean coal, quartz, and kaolinite (treated with 6 mg/L PVP).

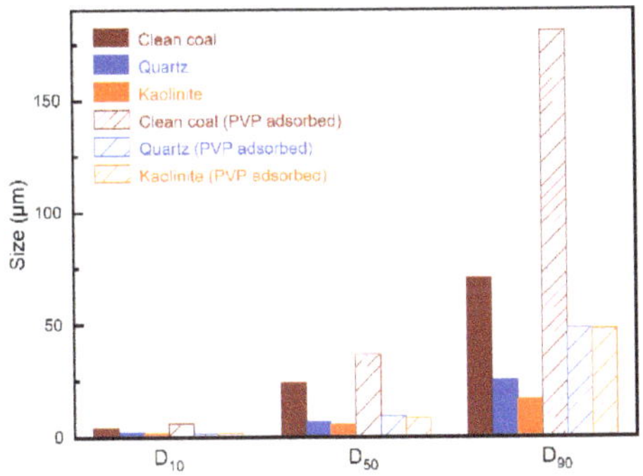

Figure 9. D10, D50, and D90 of clean coal, quartz, and kaolinite (treated with 6 mg/L PVP).

3.5. XPS

Figure 10 shows the wide energy spectrums of the clean coal, kaolinite, and quartz surfaces tested by XPS. On the one hand, the results (Figure 10a) illustrate that C (75.6%), O (19.0%), N (2.1%), Si (2.3%), and Al (1.0%) elements occupied the surface of clean coal. It was reported earlier that the proportion of N on the surface of PVP was higher than on coal [27]. Thus, the adsorption of PVP changed the elemental composition of the coal surface, mainly increasing the proportion of N from 2.1% to 3.6% (Figure 10b); on the other hand, the maximum peak of N1s also appeared on the kaolinite (Figure 10d) and quartz (Figure 10e) surfaces after PVP absorption, which can be attributed to the lactam group of PVP. The Si and Al content decreased with the absorption of PVP. The increase in the proportion of N observed on the surfaces of kaolinite and quartz was much lower than in coal [37,38]. In terms of the detection method, Uv-vis tends to analyze the adsorption amount of PVP while XPS tends to detect the surface adsorption form. Although the adsorption equilibrium amount of PVP in the gangue minerals was higher than in clean coal, XPS is a test that can be conducted for flat surfaces, excluding the influence of the grain size. In conclusion, PVP could be adsorbed on the coal, kaolinite, and quartz surfaces, resulting in better selectively in coal flotation.

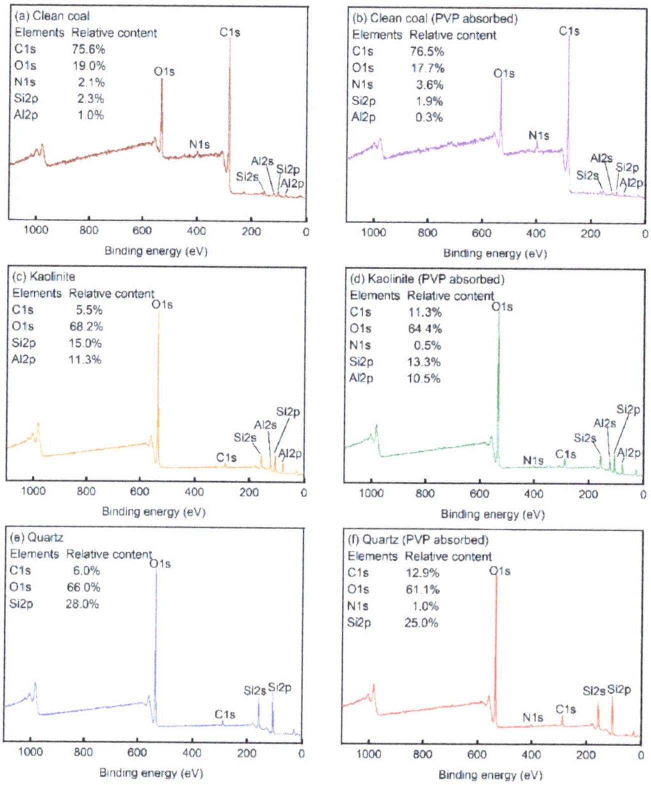

Figure 10. Changes in the wide energy spectrums of clean coal, kaolinite, and quartz tested by XPS: (**a**) clean coal, (**b**) clean coal with PVP treatment; (**c**) kaolinite, (**d**) kaolinite with PVP treatment; (**e**) quartz, (**f**) quartz with PVP treatment.

4. Conclusions

This study investigated the influence of PVP on the efficient separation of ultrafine coal. The poor flotation performance of ultrafine coal can be attributed to the initial high ash content and ultrafine grain size. PVP was used to adjust the process before flotation and improve the flotation performance. The adsorption, zeta potential, particle size distribution, and an XPS test were investigated to illustrate the mechanism of the adjusting role played by PVP. The supporting analyses and the main conclusion are summarized as follows:

(a) PVP could be adsorbed on the surface of ultrafine coal. The adsorption amount of PVP was affected by the initial concentration of PVP and the adsorption time. The adsorption amount of PVP could be used to adjust the zeta potential of the particles and influence the attraction between them.

(b) The adsorption characteristics of PVP affected the flocculation of clean coal, kaolinite, and quartz, forcing more particles to float upward. The flotation rate and combustible recovery were significantly improved, and the *SI* value revealed an increase in the selectivity at low PVP concentrations.

(c) The separation process of ultrafine particles was highly sensitive to the polymer reagents. The addition of low concentrations of PVP led to an improvement in the separation efficiency, whereas excessive amounts of PVP were not conducive to separation.

(d) Polymer agents have wide application prospects and can be used to solve problems that are not proficient for the short-chain molecule. As polymers are easy to load or modify, the influence of the long chain length of PVP can be further researched to achieve more efficient separation.

Author Contributions: Conceptualization, Y.M.; methodology, Y.M. and Q.L.; software, Y.M.; validation, C.W. and W.H.; formal analysis, W.H.; investigation, Y.M.; resources, D.C.; data curation, Q.L.; writing—original draft preparation, Y.M.; writing—review and editing, W.H.; visualization, Q.L.; supervision, K.L.; project administration, C.W. and K.L.; funding acquisition, K.L. All authors have read and agreed to the published version of the manuscript.

Funding: This research was funded by the project from the Shenzhen Development and Reform Commission (No. XMHT20190203001), Guangdong Innovative and Entrepreneurial Research Team Program (No. 2016ZT06N532) with Shenzhen Government Related Supporting Fund (No. KYT-DPT20181011104002), and RevTech-SUSTech Joint RMS Technology Development Project (OR2109006).

Acknowledgments: The authors acknowledge the assistance of SUSTech Core Research Facilities. The authors thanks for the help from Junguo, Li and Yumeng Chen during the research.

Conflicts of Interest: The authors declare no conflict of interest.

References

1. Cai, Y.; Du, M.; Wang, S.; Liu, L. Determination of oxidation properties and flotation parameters of low-rank coal slimes. *Powder Technol.* **2019**, *353*, 20–26. [CrossRef]
2. Wang, G.; Bai, X.; Wu, C.; Li, W.; Liu, K.; Kiani, A. Recent advances in the beneficiation of ultrafine coal particles. *Fuel Process. Technol.* **2018**, *178*, 104–125. [CrossRef]
3. Zhao, X.; Tang, Y.; Zhao, B.; Wu, C.; Li, J.; Chu, C. Collecting behaviors of high internal phase (HIP) emulsion in flotation of ultrafine high-ash content coal slime. *Int. J. Coal Prep. Util.* **2021**, *41*. [CrossRef]
4. Wang, J.; Wang, L. Improving column flotation of oxidized or ultrafine coal particles by changing the flow pattern of air supply. *Miner. Eng.* **2018**, *124*, 98–102. [CrossRef]
5. Norori, M.A.; Brito, P.P.R.; Hadler, K.; Cole, K.; Cilliers, J.J. The effect of particle size distribution on froth stability in flotation. *Sep. Purif. Technol.* **2017**, *184*, 240–247. [CrossRef]
6. Tao, X.; Cao, Y.; Liu, J.; Shi, K.; Liu, J.; Fan, M. Studies on characteristics and flotation of a hard-to-float high-ash fine coal. In Proceedings of the 6th International Conference on Mining Science & Technology, Xuzhou, China, 18 October 2009.
7. Akdemir, Ü.; Sönmez, I. Investigation of coal and ash recovery and entrainment in flotation. *Fuel Process. Technol.* **2003**, *82*, 1–9. [CrossRef]
8. Li, C.; Xing, Y.; Gui, X.; Cao, Y. Enhancement of oxidized coal flotation by preconditioning with positive-charged microbubbles. *Int. J. Coal Prep. Util.* **2020**, *40*, 553–563. [CrossRef]
9. Zhang, Z.; Ren, L.; Zhang, Y. Role of nanobubbles in the flotation of fine rutile particles. *Miner. Eng.* **2021**, *172*, 107140. [CrossRef]
10. Huo, X.; Zuo, W.; Shi, F.; Huang, W. Coal middling retreatment using high voltage pulse technique. Part 1: Experimental findings. *Fuel* **2022**, *314*, 123066. [CrossRef]
11. Zhou, S.; Wang, X.; Bu, X.; Wang, M.; An, B.; Shao, H. A novel flotation technique combining carrier flotation and cavitation bubbles to enhance separation efficiency of ultra-fine particles. *Ultrason. Sonochem.* **2020**, *64*, 105005. [CrossRef]
12. Capes, C.E.; Darcovich, K. A survey of oil agglomeration in wet fine coal processing. *Powder Technol.* **1984**, *40*, 43–52. [CrossRef]
13. Kadagala, M.R.; Nikkam, S.; Tripathy, S.K. A review on flotation of coal using mixed reagent systems. *Miner. Eng.* **2021**, *173*, 107217. [CrossRef]
14. Dey, S. Enhancement in hydrophobicity of low rank coal by surfactants: A critical overview. *Fuel Process. Technol.* **2012**, *94*, 151–158. [CrossRef]
15. Vamvuka, D.; Agridiotis, V. The effect of chemical reagents on lignite flotation. *Int. J. Miner. Process.* **2001**, *61*, 209–224. [CrossRef]
16. Ozer, M.; Basha, O.M.; Morsi, B. Coal-Agglomeration Processes: A Review. *Int. J. Coal Prep. Util.* **2016**, *37*, 131–167. [CrossRef]
17. Chen, Y.; Zhou, B.; Zhang, X.; Yang, S.; Huang, W. Understanding the role of kerosene on the coal particle and bubble attachment process. *Fuel* **2022**, *307*, 121915. [CrossRef]
18. Zhao, J.; Fu, X.; Wang, J.; Song, G.; Wang, K. Influence of ultra-fine grinding mode on ultra-clean coal separation effect. *J. China Coal Soc.* **2016**, *41*, 3108–3114.
19. Xia, Y.; Zhang, R.; Xing, Y.; Gui, X. Improving the adsorption of oily collector on the surface of low-rank coal during flotation using a cationic surfactant: An experimental and molecular dynamics simulation study. *Fuel* **2019**, *235*, 687–695. [CrossRef]
20. Marsalek, R.; Pospisil, J.; Taraba, B. The influence of temperature on the adsorption of CTAB on coals. *Colloids Surf. A Physicochem. Eng. Asp.* **2011**, *383*, 80–85. [CrossRef]
21. Wen, B.; Xia, W.; Sokolovic, J.M. Recent advances in effective collectors for enhancing the flotation of low rank/oxidized coals. *Powder Technol.* **2017**, *319*, 1–11. [CrossRef]

22. Zou, W.; Gong, L.; Huang, J.; Zhang, Z.; Sun, C.; Zeng, H. Adsorption of hydrophobically modified polyacrylamide P(AM-NaAA-C(16)DMAAC) on model coal and clay surfaces and the effect on selective flocculation of fine coal. *Miner. Eng.* **2019**, *142*, 105887. [CrossRef]
23. Li, C.; Xu, M.; Zhang, H. Efficient separation of high-ash fine coal by the collaboration of nanobubbles and polyaluminum chloride. *Fuel* **2020**, *260*, 116325. [CrossRef]
24. Lv, S.; Peng, W.; Cao, Y.; Liu, S.; Wang, W.; Fan, G. Synthesis and characterisation of a novel pH-sensitive flocculant and its flocculation performance. *J. Mol. Liq.* **2022**, *348*, 118480. [CrossRef]
25. Palchoudhury, S.; Lead, J.R. A Facile and Cost-Effective Method for Separation of Oil-Water Mixtures Using Polymer-Coated Iron Oxide Nanoparticles. *Environ. Sci. Technol.* **2014**, *48*, 14558–14563. [CrossRef]
26. Alabresm, A.; Chen, Y.P.; Decho, A.W.; Lead, J. A novel method for the synergistic remediation of oil-water mixtures using nanoparticles and oil-degrading bacteria. *Sci. Total Environ.* **2018**, *630*, 1292–1297. [CrossRef] [PubMed]
27. Lin, Q.; Mei, Y.; Huang, W.; Zhang, B.; Liu, K. Understanding the Role of Polyvinylpyrrolidone on Ultrafine Low-Rank Coal Flotation. *ACS Omega* **2022**, *7*, 10196–10204. [CrossRef] [PubMed]
28. Barbian, N.; Ventura, M.E.; Cilliers, J.J. Dynamic froth stability in froth flotation. *Miner. Eng.* **2003**, *16*, 1111–1116. [CrossRef]
29. Bu, X.; Wang, X.; Zhou, S.; Li, B.; Zhan, H.; Xie, G. Discrimination of Six Flotation Kinetic Models Used in the Conventional Flotation and Carrier Flotation of -74 μm Coal Fines. *ACS Omega* **2020**, *5*, 13813–13821. [CrossRef]
30. Xu, M. Modified flotation rate constant and selectivity index. *Miner. Eng.* **1998**, *11*, 271–278. [CrossRef]
31. Hu, S.; Chen, Y.; Wu, C.; Li, J.; Liu, K. The performance and dispersing mechanism of anionic dispersants in slurries prepared by upgraded coal. *Colloids Surf. A Physicochem. Eng. Asp.* **2020**, *606*, 125450. [CrossRef]
32. Zhang, Y.; Hu, S.; Yang, X.; Jiang, F.; Wu, C.; Li, J. Performance and mechanism of polyacrylamide stabilizers in coal water slurry. *Colloids Surf. A Physicochem. Eng. Asp.* **2021**, *630*, 127544. [CrossRef]
33. Li, L.; Li, Z.; Ma, C.; Wang, J.; Cao, X.; Wang, P. Molecular dynamics simulations of nonionic surfactant adsorbed on subbituminous coal model surface based on XPS analysis. *Mol. Simul.* **2019**, *45*, 736–742. [CrossRef]
34. Zhang, Q.; Niu, C.; Bu, X.; Bilal, M.; Ni, C.; Peng, Y. Enhancement of Flotation Performance of Oxidized Coal by the Mixture of Laurylamine Dipropylene Diamine and Kerosene. *Minerals* **2021**, *11*, 1271. [CrossRef]
35. Xing, Y.; Gui, X.; Karakas, F.; Cao, Y. Role of Collectors and Depressants in Mineral Flotation: A Theoretical Analysis Based on Extended DLVO Theory. *Minerals* **2017**, *7*, 223. [CrossRef]
36. Hussain, S.A.; Demirci, S.; Ozbayoglu, G. Zeta potential measurements on three clays from Turkey and effects of clays on coal flotation. *J. Colloid Interface Sci.* **1996**, *184*, 535–541. [CrossRef]
37. Sahoo, H.; Sinha, N.; Rath, S.S.; Das, B. Ionic liquids as novel quartz collectors: Insights from experiments and theory. *Chem. Eng. J.* **2015**, *273*, 46–54. [CrossRef]
38. Tian, M.; Khoso, S.A.; Wang, L.; Sun, W.; Zhang, C.; Hu, Y. Selective separation behavior and its molecular mechanism of cassiterite from quartz using cupferron as a novel flotation collector with a lower dosage of Pb^{2+} ions. *Appl. Surf. Sci.* **2019**, *486*, 228–238. [CrossRef]

minerals

MDPI

Article

Sedimentation of Fine Arsenopyrite with PEI and the Flotation Significance

Pingtian Ming [1], Dan Zou [2], Fei Li [1,*], Qingqing Xing [1] and Zhen Wang [2,*]

1 Qinghai Engineering Research Center for Gold Mineral Resource Development Dressing and Metallurgy Pilot Plant, Dulan Jin Hui Mining Limited Corporation, Haixi Zhou 816100, China; ptming4500@sina.com (P.M.); ramble77077@sina.com (Q.X.)

2 Key Laboratory of Solid Waste Treatment and Resource Recycle Ministry of Education, Southwest University of Science and Technology, Mianyang 621010, China; zoudan@mails.swust.edu.cn

* Correspondence: a376169069@163.com (F.L.); wangzhen@swust.edu.cn (Z.W.)

Abstract: The flotation of fine mineral particles is always a difficult problem. The flotation of fine arsenopyrite particles (−20 μm) in a sodium butyl xanthate (SBX) system was studied by using polyethyleneimine (PEI) as a flocculant. The flocculation properties of PEI on fine arsenopyrite were studied using sedimentation tests. The results showed that the optimum pH for the sedimentation of PEI was approximately 7.5; the higher the molecular weight (M.W.) of the flocculant, the better the sedimentation effect. In the flotation experiments, it was found that the flotation recovery of PEI-3 with high M.W. as flocculant was only 57%, while the flotation recovery of PEI-2 with medium M.W. was 90% under respective optimum conditions. The contact angle tests showed that the natural contact angle of arsenopyrite was 37°; the addition of moderate PEI-2 had a slightly negative influence on the hydrophobicity of arsenopyrite in the SBX system. From the size analysis results, the maximum particle size (D_{100}) and median size (D_{50}) of the arsenopyrite increased from 20 and 11 μm to 48 and 28 μm after treatment with 40 mg/L PEI-2, a size more conducive to bubble capture. From the combination of these results, it can be concluded that PEI-2 improved the flotation of fine arsenopyrite mainly by increasing the particle size to a suitable range through flocculation. The XPS results indicated that the adsorption of PEI-2 on the arsenopyrite surface was due to the chemisorption between the imino group and the active Fe/As sites. Applying PEI-2 to a fine disseminated arsenopyrite-type gold ore, a concentrate containing 36 g/t Au with a Au recovery of 88% can be obtained.

Keywords: flocculation; arsenopyrite; particle size; adsorption; flocculant

Citation: Ming, P.; Zou, D.; Li, F.; Xing, Q.; Wang, Z. Sedimentation of Fine Arsenopyrite with PEI and the Flotation Significance. *Minerals* **2022**, *12*, 824. https://doi.org/10.3390/min12070824

Academic Editor: Hyunjung Kim

Received: 16 May 2022
Accepted: 26 June 2022
Published: 28 June 2022

1. Introduction

The Wulonggou gold mine is located in Qinghai Province, Western China. It is one of the largest ore concentration areas in the east Kunlun metallogenic belt [1]. The ore is mainly disseminated and veined disseminated and has a brecciform, fine-grained structure. The main mineral assemblage is sericite, (white) mica, quartz, pyrite and arsenopyrite. Gold elements mainly exist in metallic sulfides, such as arsenopyrite and pyrite, in an ultra-microscopic state, and the distribution ratio of Au in arsenopyrite is approximately 70% [2]. Since the disseminated size of arsenopyrite and pyrite in ore is becoming finer, the grind fineness must be finer so that Au-bearing minerals can be liberated from gangues; now, the flotation indexes are deteriorating, highlighted by the decrease in the flotation recovery of Au (arsenopyrite) and the increase in the Au grade of tailings. The main approach under consideration at present is to seek new effective flotation reagent regimes that can improve the recovery of fine particles, as the existing flotation equipment must be adequately utilized.

Fortunately, researchers have a number of methods to make fine particles grow, by which the flotation of fines would become relatively easy [3,4]. Suitable electrolytes can make fine particles grow to large sizes through the mechanism of compressing double-layer particles (agglomeration) [5]; some long-chain polymers can make fine particles grow to large sizes through the mechanism of the bridging effect (flocculation) [6]; some hydrophobic organics can make fine particles grow to large sizes through the mechanism of the hydrophobization of the mineral surface (hydrophobic attraction) [7]. Among them, the flocculation of fine particles through long-chain polymers is the most frequently used method. The linear polymer polyacrylamide (PAM) is freely soluble in water, so it is widely applied in mining and environment industries, etc. [8]. PAM can be grafted in different groups to form cationic polyacrylamide (CPAM) or anionic polyacrylamide (APAM), by which the chemical interaction between the grafted group of PAM and mineral surface becomes possible [9]. However, the depression of such PAM-based flocculants is relatively strong, which would negatively affect the recovery of other valuable minerals. For example, a PAM-based flocculant was used in fine copper sulfide mineral flotation to improve Cu recovery, but the Mo recovery showed a significant decrease [10].

Therefore, it is important to search for new flocculants. Polyethyleneimine (PEI), a cationic flocculant, is often used to remove metal cations in wastewater through chelation, so it has the potential to be used as a flotation reagent because the interaction of a reagent with a mineral also takes place through the adsorption between the special groups of the molecule and the active sites (metal sites) on the mineral surface [11]. Here, it is employed as a new reagent in a fine arsenopyrite flotation system to mainly reflect its flocculation effect. The adsorption of PEI on the arsenopyrite surface and its effect on the flotation performance of sodium butyl xanthate (SBX) are studied through flotation tests, and the improving mechanism is also discussed.

2. Experimental Procedure

2.1. Materials

The arsenopyrite used in the experiment was obtained from Dulan Jinhui Mining Co., Ltd. Parts of the ore samples were handpicked and then ground in a ceramic ball mill. Finally, -20 μm arsenopyrite was obtained through the elutriation method and was used for sedimentation, flotation, size analysis and X-ray photoelectron spectroscopy (XPS) tests. The X-ray diffraction (XRD) spectra (Figure 1) of the samples showed that the purity of arsenopyrite met the requirement through comparison with the corresponding PDF cards in Jade (version 6.0, Materials Data, Inc., Leominster, MA, USA) [12]. The high purity of the arsenopyrite sample was also demonstrated by the chemical element analysis results of 43.46% As, 19.64% S and 32.44% Fe, which were very close to the stoichiometric ratio of arsenopyrite. A massive arsenopyrite was cut into a 3 cm $\times$ 3 cm $\times$ 1 cm cube, with one of the 3 cm $\times$ 3 cm surfaces being polished using an automatic target surface processor (EM TXP, Leica Microsystems, Wetzlar, Germany), which was used for the contact angle measurement.

NaOH and H_2SO_4 stock solutions were used as the pH regulators, except for bench scale flotation, where Na_2CO_3 was the pH modifier. Sodium butyl xanthate (SBX) was used as the collector in the micro-flotation tests, while in the bench-scale flotation of the real ore, s mixture of ammonium dibutyl dithiophosphate (ADD) and SBX was employed. PEIs with different M.W. (PEI-1, 3500; PEI-2, 10,000; PEI-3, 50,000) were used as the flocculants. Methyl isobutyl carbinol (MIBC) was used as the frother for all the flotation tests. The monomer molecular structure of PEIs is shown in Figure 2. Deionized water (18.25 MΩ·cm) was used for pure mineral tests; tap water was used for bench-scale flotation tests.

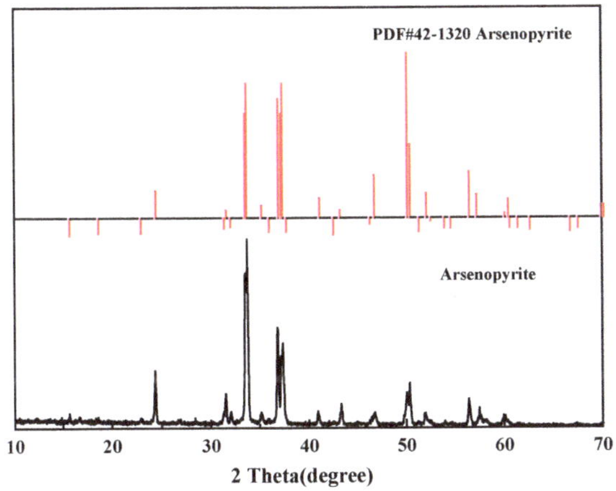

Figure 1. X-ray diffraction patterns of arsenopyrite with corresponding PDF cards.

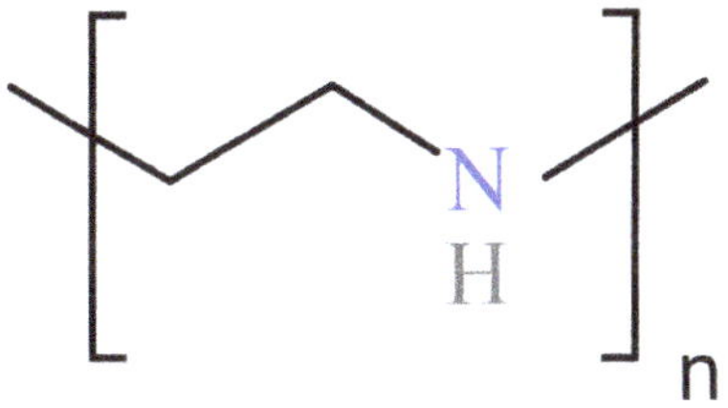

Figure 2. Monomer structure of linear polyethyleneimine.

2.2. Sedimentation Tests

Sedimentation characteristics could be used to understand the flocculation properties of fine arsenopyrite under different flocculation conditions. In the sedimentation tests of arsenopyrite, 5 g of powder, together with 90 mL of deionized water, was added to a 100 mL beaker. After adjusting the slurry pH to the desired value, the corresponding reagent(s) were added. Each interaction time was 2 min for the reagent with arsenopyrite particles under magnetic agitation. Then, the slurry was transferred to a 100 mL graduated cylinder and agitated vigorously to remain fully suspended. After the stirring was stopped, the settling behavior of the slurries was observed after a specified time by recording the solid–liquid interface. The fraction over the interface was then pored off, and the remaining part was filtered, dried and weighed to calculate the sedimentation yield [13].

2.3. Flotation

2.3.1. Micro-Flotation

Floatability tests were conducted in a 60 mL micro-flotation cell. Pure mineral particles (2.0 g) were placed in a plexiglass cell for 2 min to be conditioned, and then, a pH regulator was added to adjust it to the required pH. Under this desired pH, PEI flocculant was added, followed by SBX collector, with 3 min of stirring time after each addition. After all of the above was finished, the flotation process was performed for 4 min. After being filtered and dried, the froth products together with the tails were separately weighed. Then, the recovery was calculated in accordance with the dry weight of the products (Equation (1)). The researchers established three flotation tests in the same conditions to report the average values of the three methods.

$$R = W_C/(W_C + W_T) \quad (1)$$

where Wc is the weight of the concentrate and W_T is that of tailings.

2.3.2. Bench Scale Flotation

The bench-scale flotation experiments of the actual ore were carried out in an XFD-63 flotation cell, with 1500 mL rougher–scavenger flotation and 500 mL cleaner flotation. The impeller speed was 1650 rpm. Froth products were collected using an automatic froth scraping device. In the flotation stage, the required reagents were successively added to the flotation cell with an interval of 3 min and a flotation time of 6 min. The final products of concentrate and tailings were filtered, dried, weighed and analyzed with Au. The bench-scale locking cycle test process is shown in Figure 3. The experimental system of the real ore used the SBX + ADD as the collector.

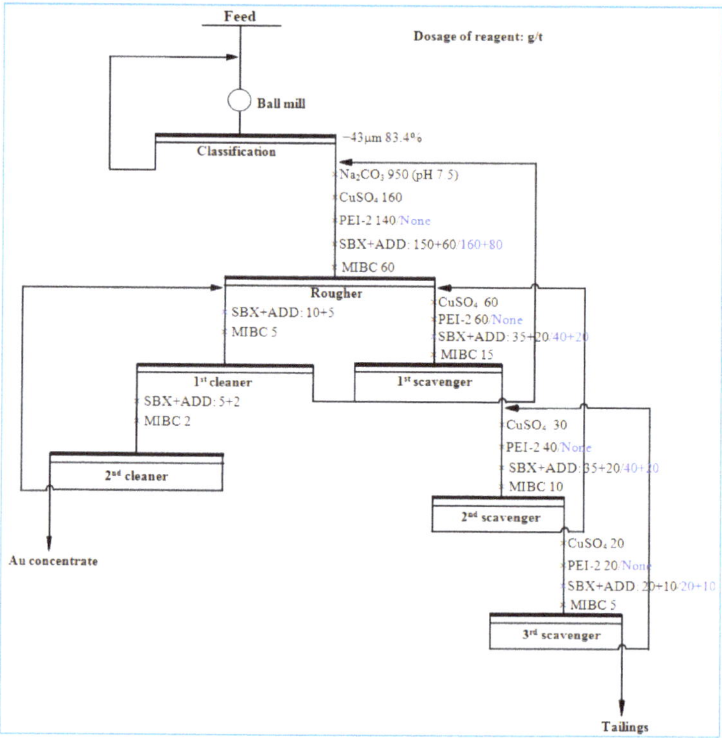

Figure 3. Flowsheet and corresponding conditions of bench-scale locked cycle tests.

2.4. *Particle Size Distribution Analysis*

Particle size distribution could directly reflect the flocculation behaviors between arsenopyrite and PEI-2. The −20 µm samples were treated with different concentrations of PEI-2 for 3 min to obtain the flocculated suspension. The laser-based particle size analysis was carried out on the treated and untreated samples with a Master-size 2000 instrument (Malvern Panalytical Ltd., Malvern, UK). Ultrasonics were not applied to all the suspensions to prevent damage to the flocs.

2.5. *Contact Angle Measurement*

The contact angle was measured using a Rame-Hart goniometer (Model 590, Rame-Hart, Succasunna, NJ, USA) through the sessile drop method [14]. The cube arsenopyrite (3 cm × 3 cm × 1 cm cube with one polished 3 cm × 3 cm surface) was conditioned with different reagents by immersing it in the corresponding reagent solution for 3 min. After, the washing and drying of the tablets were conducted. The contact angle value was determined by recording the image of the prepared sample with the water droplet. For each specific measurement, at least three tests were conducted at different locations on the surface. The reported value was the average of these measurements.

2.6. XPS Detection

A 1.0 g sample of −20 μm arsenopyrite powder was added into the 60 mL micro-flotation cell together with 40 mL distilled water. The pH was adjusted to 7.5, upon which the desired amount of PEI-2 was added and stirred for 10 min at 1900 r/min. After filtration and vacuum drying, the XPS spectra for arsenopyrite samples treated with or without PEI-2 were recorded using a K-Alpha 1063 spectrometer (Thermo Scientific Co., Waltham, MA, USA), which employs Al Kα as a sputtering source at 12 kV and 6 mA with 1.0×10^{-9} Pa pressure in the analytical chamber. The curve was fitted using XPSPEAK 4.1 software (version 4.1, Taiwan, China).

3. Results and Discussion

3.1. Sedimentation Tests

PEI dosage, pH and sedimentation time were the key factors affecting the settling. The influence of these factors on arsenopyrite (−20 μm) settlement was studied using systematic sedimentation tests. The experimental results are shown in Figure 4. As could be seen from Figure 4a, except for PEI-1, the sedimentation efficiency increased with the increase in the dosage of flocculant. According to the curve trend, PEI-3 had the best effect, with the highest sedimentation yield of 96% when the dosage was 200 mg/L. When the dosage surpassed 120 mg/L, the sedimentation yield increased slowly, and it was selected for further tests.

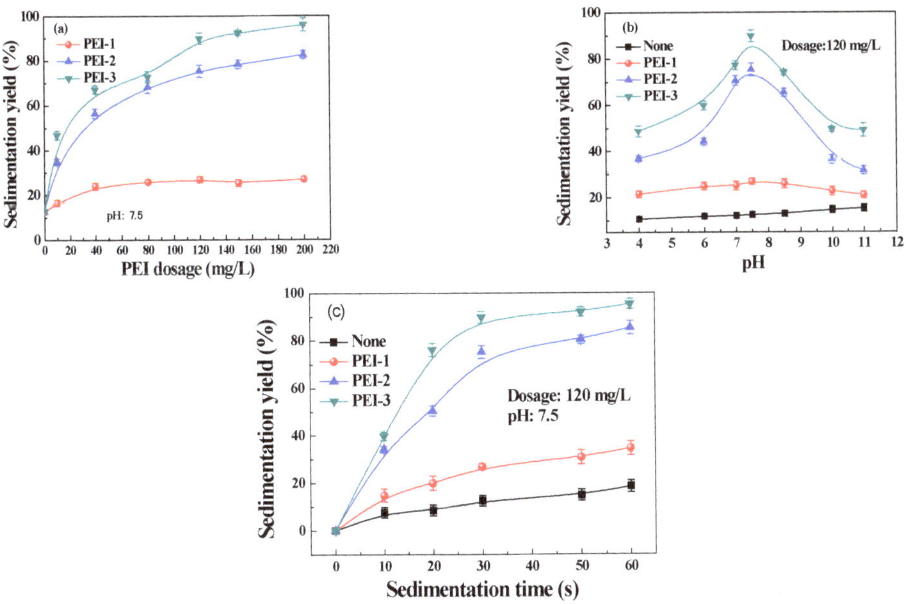

Figure 4. Effects of PEI dosage (pH = 7.5; sedimentation time 30 s) (**a**), pH (dosage, 120 mg/L; sedimentation time 30 s) (**b**) and sedimentation time (pH = 7.5; dosage, 120 mg/L) (**c**) on sedimentation yield.

The influence of pH on PEI sedimentation efficiency is shown in Figure 4b. The sedimentation efficiency of minerals in PEI-2 and PEI-3 systems increased first and then decreased with the increase in pH. The optimal value of PEI-3 was 90%, while the sedimentation yield without flocculant was only 12.6%. When pH exceeded 7.5, the decrease in sedimentation efficiency may have been caused by the decationization of the PEIs under alkaline conditions [15,16]. Therefore, the pH of 7.5 was determined as the optimum condition for the sedimentation tests.

If the sedimentation time was set as the independent variable, it was found that regardless of PEIs, the sedimentation efficiency increased with the increase in time (Figure 4c). When the sedimentation time was 60 s, the sedimentation yield was only 18.66% without flocculant, but it was as high as 95% with the addition of PEI-3.

In conclusion, under the same conditions, the higher the molecular weight of the flocculant, the better the sedimentation efficiency. In flotation experiments, however, a different result was achieved.

3.2. Flotation Tests

3.2.1. Micro-Flotation

A single mineral flotation test was conducted to explore the effects of PEI dosage, pH and SBX concentration on the flotation recovery of arsenopyrite with SBX as the collector, and the results are shown in Figure 5. As shown in Figure 5a, without PEI, the flotation recovery of arsenopyrite was lower than 50% and decreased with the increasing pH, especially when the pH was higher than 7.5. The addition of PEI-1 slightly improved the recovery of arsenopyrite in the pH range of 4–11, while the addition of PEI-2 and PEI-3 significantly improved the flotation recovery of arsenopyrite. At pulp pH 7.5, the recovery of arsenopyrite was only 44% without flocculant, while the optimum recovery increased to 80% with 120 mg/L PEI-2 as the flocculant in 8×10^{-5} M SBX solution.

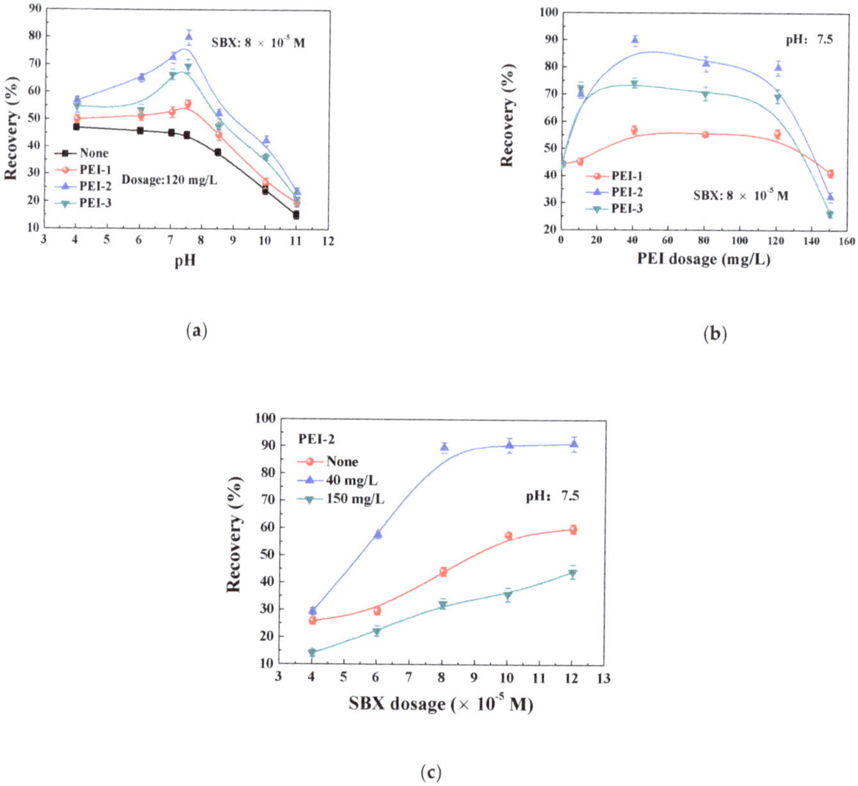

Figure 5. Effects of (**a**) pH (PEI, 120 mg/L; SBX, 8×10^{-5} M), (**b**) PEI dosage (pH, 7.5; SBX, 8×10^{-5} M) and (**c**) SBX dosage (pH, 7.5) on the recovery of arsenopyrite.

Fixing the pH to 7.5, the influence of the PEIs dosage on mineral recovery was investigated. As illustrated in Figure 5b, with the increasing PEI dosage, the flotation recovery increased first and then declined. The best dosage of PEI was 40 mg/L at the conditions of pulp pH 7.5 and SBX dosage 8×10^{-5} M. When the dosage of flocculant exceeded 120 mg/L, the recovery of arsenopyrite decreased. Evidently, adding moderate flocculant could improve the flotation of fine arsenopyrite particles, which was consistent with the previous study [17]. Additionally, in the two flotation tests, moderate PEI-2 showed the best improvement in the flotation of arsenopyrite.

With different dosages of PEI-2, the influence of SBX dosage on the recovery of arsenopyrite was researched, and the results are shown in Figure 5c. The recovery increased with the increasing SBX dosages. When the dosage of PEI-2 was 40 mg/L, its addition improved arsenopyrite flotation, while with 150 mg/L PEI-2, the recovery was even lower than that without additives, which was ascribed to the excessive flocculant, resulting in a particle agglomeration size that is too large, thus causing bubble rupture [18].

In the settlement experiment, the best settlement effect was PEI-3 (maximal M.W.). However, in the micro-flotation experiment, the best flotation recovery of arsenopyrite was PEI-2 (medium M.W.). The reason for this phenomenon was that the high molecular weight of PEI-3 resulted in the optimal flocculation ability and made the particle size of the flocculated fine particles of arsenopyrite too large, which would increase the probability of particle desorption from the bubble [19].

3.2.2. Bench-Scale Flotation

Based on the results of micro-flotation tests and the subsequent flotation condition tests on the real ore, locked cycle testing was carried out at batch scale using SBX + ADD as a mixed collector in weakly alkaline pulp (pH 7.5). The closed flotation circuit included one roughing, three scavenging, and two cleaning steps (shown in Figure 3), and all the reagent dosages and grinding fineness presented are the optimum values. The flotation results achieved with and without PEI-2 are displayed in Table 1. A concentrate containing 31.44 g/t Au with the Au recovery of 77% was achieved without PEI-2 in the flotation process at pH 7.5. When PEI-2 was added into the pulp before collectors in the roughing and scavenging stage, a concentrate with the Au grade of 36 g/t and Au recovery of 88% was obtained. For the fine-grained disseminated refractory gold ore, the flocculant PEI-2 improved the flotation of the gold-bearing arsenopyrite-type ore, and with PEI-2, a higher grade and recovery of gold concentrate could be obtained in the weak alkaline pulp. It is of great industrial application potential in fine particle flotation.

Table 1. Values of closed flotation circuit.

System	Products	Ratio (w/%)	Au Grade (g/t)	Au Recovery (%)
Without PEI-2	Au concentrate	5.39	31.44	77.28
	Tailing	94.61	0.53	22.72
	Feed	100	2.19	100
With PEI-2	Au concentrate	5.35	35.99	87.92
	Tailing	94.65	0.28	12.08
	Feed	100	2.19	100

3.3. *Size Analysis Results*

The addition of PEI-2 could increase the particle size of fine arsenopyrite according to the results of sedimentation tests. Here, the size testing was conducted on the particles treated with 0, 40 and 150 mg/L PEI-2 using a laser particle sizer.

The results shown in Figure 6 provide intuitive data. The maximum particle size (D_{100}) and median size (D_{50}) of the arsenopyrite without PEI treatment were 20 and 11 μm, which were far below the sizes that can be effectively floated [20], while the corresponding data increased to 48 and 28 μm after treatment with 40 mg/L PEI-2. This increase in particle size was responsible for the improvement in sedimentation yield (Figure 4) and flotation recovery (Figure 5). When the PEI-2 dosage was 150 mg/L, the flocculated arsenopyrite suspension presented the D_{100} and D_{50} of 116 and 69 µm, respectively. These values increased to approximately six times those of the original ones, which resulted in a greater sedimentation yield but lower flotation recovery [21]. These size analysis results further demonstrated the conclusion regarding the size increase according to the sedimentation tests.

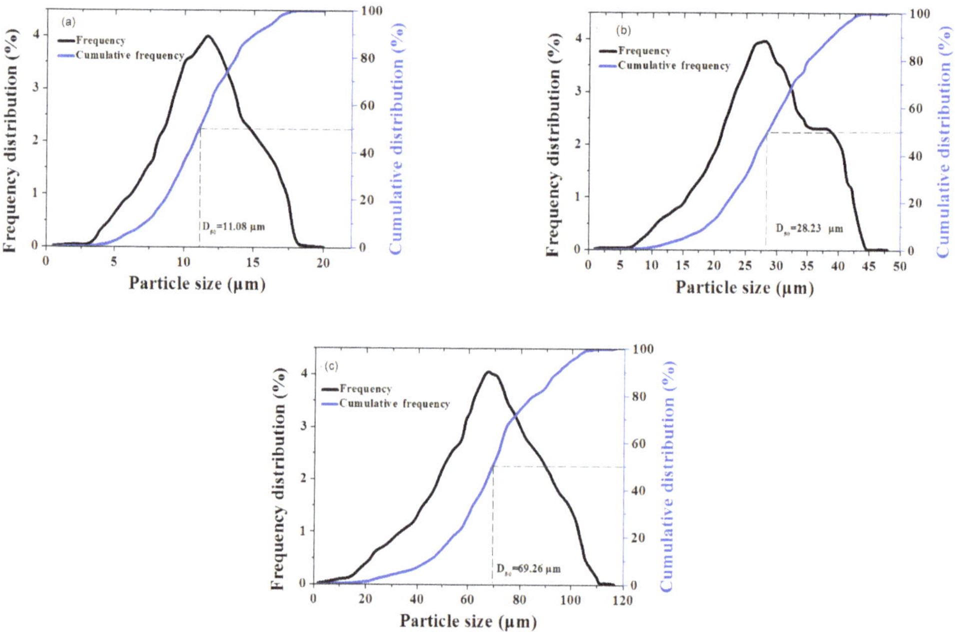

Figure 6. Size distribution of the arsenopyrite particles treated with (**a**) 0, (**b**) 40 and (**c**) 150 mg/L PEI-2. (Dotted line represents 50% cumulative yield corresponded particle size.)

3.4. Contact Angle Measurement

Enhancing the recovery of fines can be accomplished by increasing the particle size to a suitable range or improving the hydrophobicity of the particle surface [3]. The contact angles of the arsenopyrite surface under different conditions were measured, which, combined with the size analysis results, are shown in Figure 6 to determine the relationships between the good flotation recovery and the two factors.

The contact angle results are shown in Figure 7. The natural contact angle of arsenopyrite was 36°, which is close to some of those reported in the literature [22]. When it was treated with distilled water of pH 7.5, the contact angle decreased slightly, mainly due to the formation of hydrophilic metal hydroxyl compounds [23]. For the surface treated with 8×10^{-5} M SBX (pH 7.5), the contact angle shifted to 62°, showing that SBX could greatly improve the hydrophobicity of arsenopyrite. At this condition, the recovery was only 44%, as shown in Figure 5a. However, when the arsenopyrite surface was treated with 40 mg/L PEI-2 solution, the contact angle dropped to 28°, indicating that it resulted in a decrease in the hydrophobicity of the surface due to its hydrophilicity. When the surface was treated with 40 mg/L PEI-2 solution first and 8×10^{-5} M SBX after, the contact angle of 57° was obtained, slightly lower than that with a single SBX. However, the flotation recovery was 79% under this condition (Figure 5). Therefore, the improvement of PEI-2 upon arsenopyrite flotation in the SBX system was not due to increasing the hydrophobicity of the arsenopyrite surface but due to increasing the particle size to a suitable range via the flocculation effect.

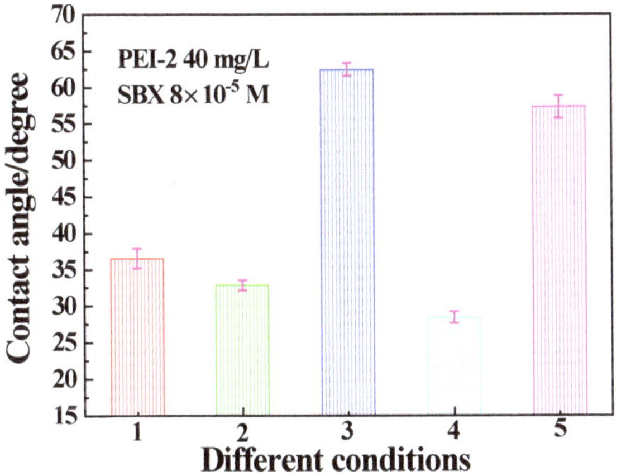

Figure 7. Contact angle measurement results under different conditions (1—natural; 2—treated with aqueous solution at pH 7.5; 3—treated with 8×10^{-5} M SBX at pH 7.5; 4—treated with 40 mg/L PEI-2 at pH 7.5; 5—treated with 40 mg/L PEI-2 then 8×10^{-5} M SBX, at pH 7.5).

3.5. XPS Analysis

As shown by the size analysis and contact angle results, with moderate PEI-2 flocculant (40 mg/L), the arsenopyrite size had a modest increase (D_{50} 28 μm), while the contact angle had only a very mild decrease in the SBX system. This is the reason for the improvement in PEI-2 upon the xanthate flotation of fine arsenopyrite. Studies on the adsorption interaction between SBX and sulfides are relatively abundant [24], but those on the PEI/arsenopyrite are few. Therefore, the XPS data of the arsenopyrite surface before and after treatment with PEI-2 were detected and analyzed. The As 3d and Fe 2p spectra are shown in Figure 8.

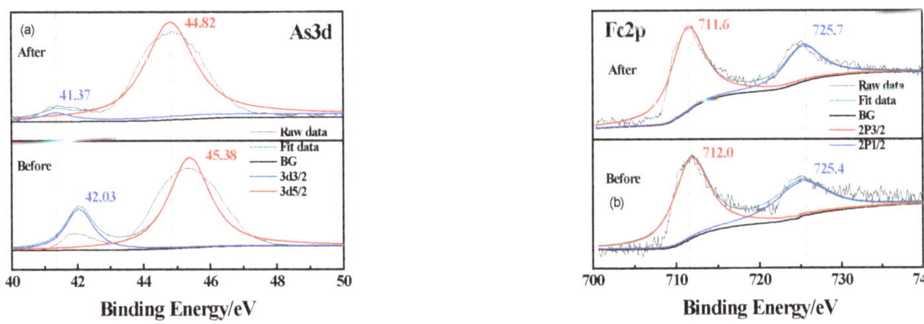

Figure 8. The (**a**) As 3d and (**b**) Fe 2p spectrum before and after interaction with PEI-2.

According to previous reports, the As and Fe atoms on the arsenopyrite surface were all likely to be the active sites for flotation reagent adsorption [25]. Without PEI interaction, the As 3d 3/2, As 3d 5/2, Fe 2p 1/2 and Fe 2p 3/2 peaks occurred at 41.37, 44.82, 711.6 and 725.7 eV, respectively. The corresponding BE shifted by 0.66, 0.56, 0.4 and −0.3 eV (all over the instrumental error of 0.2 eV), indicating the occurrence of chemical changes in the Fe and As active sites, i.e., the chemisorption of PEI-2 on the arsenopyrite surface. In the treatment of wastewater containing heavy metal ions, PEIs have been used as flocculants to collect these cations through coordination with the imino groups [26]. The minerophilic group of PEIs is the imino group. Different imino groups in the polymer can coordinate with the Fe/As atoms on the mineral surface, and some of the remaining imino groups are adsorbed with the active sites into other particles, leading to the bridging flocculation of the fine arsenopyrite particles.

3.6. Interaction Mechanism Model

Figure 9 shows the possible mechanism of flocculant PEI-2 in the flotation of fine arsenopyrite (−20 μm, D_{50} 11 μm) with collector SBX. The mineral particles with small sizes had a low collision probability with bubbles and thus had a low recovery (Figure 5), which was the reason to consider adding flocculants. When PEI-2 was added, these fines would form bigger particles through the bridging action of PEI-2 molecules. The adsorption of PEI-2 on the arsenopyrite surface was due to the coordination adsorption between the imino group and the active Fe/As sites. Different agglomerations can also further grow via the hydrophobic force between the hydrophobic group of different PEI-2 molecules. These resulted in an appropriate particle (agglomeration, D_{50} 28.23 μm) size for flotation. The hydrophobicity of the arsenopyrite surface did not reduce to a large extent (Figure 6), so it still had enough active sites for the collector SBX adsorption. Hence, the flotation of fine arsenopyrite in the SBX system could be improved by the addition of PEI-2 (Figure 5).

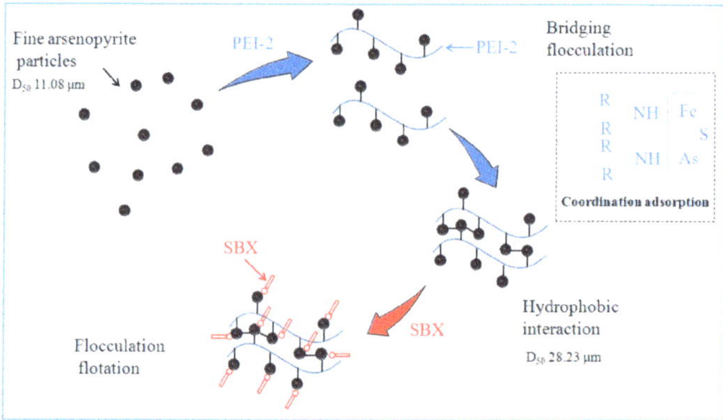

Figure 9. The possible flocculation mechanism of PEI-2 improved the flotation system of arsenopyrite.

4. Conclusions

In the xanthate system, arsenopyrite with a small particle size (−20 μm) had a poor flotation effect. This could be improved by adding moderate flocculant PEI-2, which effectively increased the particle size from D_{50} 11 μm to 28 μm at optimum dosage. Therefore, it improved the flotation of fines mainly by increasing the particle size to a suitable range through flocculation. The adsorption of PEI-2 on the arsenopyrite surface was through the coordination adsorption between the imino group and the active Fe/As sites. For a disseminated arsenopyrite-type gold ore, with the application of PEI-2, a better concentrate containing 36 g/t Au with Au recovery of 88% can be obtained with PEI-2.

Author Contributions: Conceptualization, P.M.; data curation, D.Z.; formal analysis, F.L.; funding acquisition, Z.W.; investigation, Q.X. All authors have read and agreed to the published version of the manuscript.

Funding: The Doctoral Foundation of Southwest University of Science and Technology (No. 20ZX7149) and The Sichuan Science and Technology Program of China (No. 2021YFG0269; 2022YFQ0074; 2022YFS0453).

Data Availability Statement: Not Applicable.

Acknowledgments: The authors acknowledge the support of the Doctoral Foundation of Southwest University of Science and Technology, China (no. 20ZX7149), and the Sichuan Science and Technology Program of China (no. 2021YFG0269; 2022YFQ0074; 2022YFS0453).

Conflicts of Interest: The authors declare no conflict of interest.

References

1. Sun, D.; Li, H.; Jiang, M.; Wang, P.; Luo, Y.; Liu, Z. Applications of time domain airborne electromagnetic and aeromagnetic in Wulonggou gold deposit exploration area. *Prog. Geophys.* **2017**, *32*, 2533–2544. (In Chinese) [CrossRef]
2. Wang, X.; Zhao, K.; Bo, H.; Yan, W.; Wang, Z.; Gu, G.; Gao, Z. Improved flotation of auriferous arsenopyrite by using a novel mixed collector in weakly alkaline pulp. *Physicochem. Probl. Miner. Process.* **2020**, *56*, 996–1004. [CrossRef]
3. Farrokhpay, S.; Filippov, L.; Fornasiero, D. Flotation of fine particles: A review. *Miner. Process. Extract. Metall. Rev.* **2021**, *42*, 473–483. [CrossRef]
4. Yin, W.Z.; Wang, J.Z. Effects of particle size and particle interactions on scheelite flotation. *Trans. Nonferrous Met. Soc. China* **2014**, *24*, 3682–3687. [CrossRef]
5. Chakladar, S.; Banerjee, R.; Mohanty, A.; Chakravarty, S.; Patar, P.K. Turpentine oil: A novel and natural bridging liquid for agglomeration of coal fines of high ash coals. *Int. J. Coal Prep. Util.* **2020**, 1–13. [CrossRef]
6. Hornn, V.; Ito, M.; Shimada, H.; Tabelin, C.B.; Jeon, S.; Park, I.; Hiroyoshi, N. Agglomeration-flotation of finely ground chalcopyrite and quartz: Effects of agitation strength during agglomeration using emulsified oil on chalcopyrite. *Minerals* **2020**, *10*, 380. [CrossRef]
7. Yang, B.; Song, S. Hydrophobic agglomeration of mineral fines in aqueous suspensions and its application in flotation: A review. *Surf. Rev. Lett.* **2014**, *21*, 1430003. [CrossRef]
8. Molaei, N.; Shoaib, M.; Forster, J.; Khan, S.; Wani, O.B.; Bobicki, E.R. Surface interaction between phyllosilicate particles and sustainable polymers in flotation and flocculation. *RSC Adv.* **2022**, *12*, 3708–3715. [CrossRef]
9. Estrada, D.; Echeverry, L.; Ramirez, A.; Gutierrez, L. Molybdenite flotation in the presence of a polyacrylamide of low anionicity subjected to different conditions of mechanical shearing. *Minerals* **2020**, *10*, 895. [CrossRef]
10. Castro, S.; Laskowski, J.S. Depressing effect of flocculants on molybdenite flotation. *Miner. Eng.* **2015**, *74*, 13–19. [CrossRef]
11. Wang, G.; Zhang, Y.; Jiang, S.; Ma, X.; Wei, B. Removal and recovery of cobalt from Co (II)–containing water samples by dithiocarboxyl polyethyleneimine. *Sep. Purif. Technol.* **2020**, *251*, 117338. [CrossRef]
12. Liu, N.; Wang, Z.; Xiao, J.; Wang, H.; Deng, B.; Zhang, Y.; Chen, C. Novel selective depressant of titanaugite and implication for ilmenite flotation. *Minerals* **2019**, *9*, 703. [CrossRef]
13. Jeldres, R.I.; Piceros, E.C.; Wong, L.S.; Leiva, W.H.; Herrera, N.; Toledo, P.G. Dynamic moduli of flocculated kaolinite sediments: Effect of salinity, flocculant dose, and settling time. *Colloid Polym. Sci.* **2018**, *296*, 1935–1943. [CrossRef]
14. Dalton, L.E.; Tapriyal, D.; Crandall, D.; Goodman, A.; Shi, F.; Haeri, F. Contact angle measurements using sessile drop and micro-CT data from six sandstones. *Transport. Porous Med.* **2020**, *133*, 71–83. [CrossRef]
15. Kashani, A.H.N.; Rashchi, F. Separation of oxidized zinc minerals from tailings: Influence of flotation reagents. *Miner. Eng.* **2008**, *21*, 967–972. [CrossRef]
16. Dutta, K.; Das, R.; Medeiros, J.; Kanjilal, P.; Thayumanavan, S. Charge-Conversion Strategies for Nucleic Acid Delivery. *Adv. Funct. Mater.* **2021**, *31*, 2011103. [CrossRef]
17. Zhang, X.; Gu, X.; Han, Y.; Parra-Álvarez, N.; Claremboux, V.; Kawatra, S.K. Flotation of iron ores: A review. *Miner. Process. Extract. Metall. Rev.* **2021**, *42*, 184–212. [CrossRef]
18. Deng, Y.; Sabatier, F.; Dewil, R.; Flamant, G.; Le Gal, A.; Gueguen, R.; Ansart, R. Dense upflow fluidized bed (DUFB) solar receivers of high aspect ratio: Different fluidization modes through inserting bubble rupture promoters. *Chem. Eng. J.* **2021**, *418*, 129376. [CrossRef]
19. Liu, P.; Qin, Y.; Liu, S.; Hao, Y. Non-linear gas desorption and transport behavior in coal matrix: Experiments and numerical modeling. *Fuel* **2018**, *214*, 1–13. [CrossRef]
20. Bu, X.; Xie, G.; Chen, Y.; Ni, C. The order of kinetic models in coal fines flotation. *Int. J. Coal Prep. Util.* **2017**, *37*, 113–123. [CrossRef]
21. Senior, G.D.; Shannon, L.K.; Trahar, W.J. The flotation of pentlandite from pyrrhotite with particular reference to the effects of particle size. *Int. J. Miner. Process.* **1994**, *42*, 169–190. [CrossRef]
22. Qian, W.E.I.; Dong, L.Y.; Yang, C.R.; Liu, X.D.; Fen, J.I.A.O.; Qin, W.Q. Selective depression mechanism of combination of lime and sodium humate on arsenopyrite in flotation separation of Zn–As bulk concentrate. *Trans. Nonferrous Met. Soc. China* **2022**, *32*, 668–681.
23. Hu, Y.; Wu, M.; Liu, R.; Sun, W. A review on the electrochemistry of galena flotation. *Miner. Eng.* **2020**, *150*, 106272. [CrossRef]
24. Wang, Z.; Cao, J.; Wang, L.; Xiao, J.; Wang, J. Selective depression of arsenopyrite with in situ generated nanoparticles in pyrite flotation. *Miner. Eng.* **2021**, *173*, 107223. [CrossRef]
25. Li, Y.Q.; He, Q.; Chen, J.H.; Zhao, C.H. Electronic and chemical structures of pyrite and arsenopyrite. *Mineral. Mag.* **2015**, *79*, 1779–1789. [CrossRef]
26. Gao, B.; An, F.; Liu, K. Studies on chelating adsorption properties of novel composite material polyethyleneimine/silica gel for heavy-metal ions. *Appl. Surf. Sci.* **2006**, *253*, 1946–1952. [CrossRef]

minerals

Article

Flocculation Behavior of Ultrafine Silica Particles in Acid Leaching Pulp by Nonionic Polymeric Flocculants

Bao Guo [1,*], Xinlei Zhan [1], Kaixi Jiang [1,2,3,*], Hongzhen Xie [2] and Rongdong Deng [1]

[1] Zijin School of Mining and Geology, Fuzhou University, Fuzhou 350108, China
[2] State Key Laboratory of Comprehensive Utilization of Low Grade Refractory Gold Ores, Zijin Mining Group, Longyan 364200, China
[3] BGRIMM Technology Group, Beijing 100070, China
* Correspondence: guobao@fzu.edu.cn (B.G.); jiangkx@bgrimm.com (K.J.)

Abstract: Sedimentation of ultrafine silica particles that exist in acid leaching pulp and their separation from Pregnant Leach Solution largely determines the efficiency of a hydrometallurgical process utilizing copper oxide ore. Thickener on a larger scale can allow longer sedimentation, generating low overflow turbidity but high economic input. In this paper, the flocculation behavior of quartz particles in sulfuric acid solution using nonionic flocculants polyethylene oxide (PEO) and polyacrylamide (PAM), as well as ionic cofactor montmorillonite (MMT) and nonionic cofactor tannic acid (TA), were investigated, with the dynamic size of flocs and counts of fines being monitored using an in situ particle size measurement technique, namely the focused beam reflectance measurement (FBRM), under turbulent conditions. Attention was paid to variables affecting quartz flocculation properties from both physicochemical and hydrodynamic aspects such as shear intensity. The flocculation mechanism was investigated using zeta potential and dynamic light scattering measurements. It was found that the TA promotes the bridging flocculation of PEO-quartz by forming associative complexes with larger clusters in solution, while MMT electrostatically adsorbs on the quartz surface, enhancing its bridging with PAM. Low turbidity benefited from the higher shear resistance of the compact flocs structure provided by PEO/PEO + TA/PAM + MMT. Efficient solid–liquid separation was achieved by using the synergistic flocculation of small molecule cofactors and polymer flocculants.

Keywords: acid leaching; thickening; quartz; flocculation

Citation: Guo, B.; Zhan, X.; Jiang, K.; Xie, H.; Deng, R. Flocculation Behavior of Ultrafine Silica Particles in Acid Leaching Pulp by Nonionic Polymeric Flocculants. *Minerals* **2023**, *13*, 582. https://doi.org/10.3390/min13040582

Academic Editor: Kenneth N. Han

Received: 11 March 2023
Revised: 19 April 2023
Accepted: 19 April 2023
Published: 21 April 2023

1. Introduction

Solid–liquid separation in mineral processing/hydrometallurgy/wastewater treatment is frequently performed using gravity sedimentation. In the typical Solvent Extraction-Electrowinning (SX-EW) process utilizing oxidized copper ore, the Pregnant Leach Solution (PLS) is usually acquired by dense sedimentation of the acid leaching pulp, including thickening and the efficient Counter Current Decantation (CCD) [1]. However, the decomposition of silicate minerals introduces colloidal silica formed by the disordered condensation, seriously deteriorating solid–liquid separation. Kazadi found that the silica content of PLS can be minimized by controlling silica solubilization and polymerization, but the economic viability of the process still needs to be ascertained [2]. In addition, the easily argillized silicate minerals with low hardness, especially quartz, may generate a large fraction of ultrafine particles. Ultrafine particles settle slowly and incompletely, and their entrainment into the overflow can lead to the formation of third-phase crud during solvent extraction [3,4]. Therefore, polymeric flocculants with high molecule weight, typically polyacrylamide, are used to aggregate small, suspended particles into larger flocs for more efficient sedimentation. The electrostatic forces between particles are responsible mostly for repulsion and, consequently, suspension. Aggregation of fine particles can be achieved by neutralizing the electrical charge of the interacting particles, which is coagulation, in addition to bridging the particles by nonionic polymeric flocculants, which is flocculation [5–7].

High-molecule-weight flocculants can be extremely effective in promoting floc growth in previously stabilized suspensions [8]. Ionic polymeric flocculants play the dual roles of coagulation and flocculation.

In acid leaching pulp, most of the minerals' particle surfaces are positively charged as a result of protonation. By adding coagulants to the isoelectric point, usually by adjusting the solution pH, the suspension system can be destabilized. Zinc hydrometallurgy benefits from the neutralization of acid leaching pulp to pH 5.0~5.2, where coagulation occurs between silica particles and iron hydroxide particles, the solution purification method before the electrolysis process [9,10]. By using the "neutralization-coagulation" strategy, researchers attempted to remove the crud silica/clay particles in a raffinate solution, which is recirculated to the leaching process in copper hydrometallurgy [11]. However, the problem of fine particles dragged from the feed still exists. Solvent extraction as a downstream circuit operation in acidic media prevents neutralization afterwards. Cationic surfactant cetyl trimethyl ammonium bromide addition could disturb the stability of kaolinite suspensions to some extent in copper hydrometallurgy conditions, as a result of surface charge neutralization [12]. Nonionic TX 100 surfactants were also found to improve the settling rates of kaolinite in combination with nonionic polyacrylamide due to the mutual interaction of polymer and surfactants [13]. Polystyrene sulfonate was also suitable for charge neutralizing in polyethylene-oxide-induced flocculation [14]. However, the application of surfactants usually requires either excessive dosage levels or tedious operation to avoid overcharging.

In this paper, we report an operative combination of high-molecular-weight polymer flocculants and low-molecular-weight cofactors to improve the sedimentation of acid leaching pulp. The small molecules not only play the role of coagulants, but also enhance the functionality of flocculants, borrowing the idea of the flocculation cofactor that has been extensively investigated in the papermaking industry [15,16]. The performance of nonionic flocculants polyethylene oxide (PEO) and polyacrylamide (PAM), ionic cofactors montmorillonite (MMT), and tannic acid (TA) were studied in quartz–sulfuric acid system, and different flocculation models were proposed.

2. Experimental

2.1. Materials

Quartz is typically a major component within residues from the digestion of copper oxide feeds. Herein, quartz was used as a synthetic substrate to represent leaching residue in this work, the source varying in terms of the particle size. One source of quartz sample was the product of closed-circuit grinding, and its particle size distribution was measured using a Malvern Mastersizer 2000, as shown in Figure 1a. The ground sample had a D_{90} of 54.690 μm and D_{50} of 19.588 μm, while 31.98% of the mass was attributed to fine particles ($\leq$10 μm). The deviation from the Normal Distribution indicated that significant argillization had occurred, raising a problem for solid–liquid separation. As this study sought to emphasize the impact of the flocculation efficiency of fine particles, a quartz sample source of 2000 mesh was also used. The X-ray diffraction in Figure 1b confirmed the crystalline phase of the quartz samples.

The molecular structures of nonionic flocculants polyethylene oxide (PEO, from Macklin) and polyacrylamide (PAM, commercially provided as 6003s from SNF) as well as cofactor tannic acid (TA, from Aladdin) are shown in Figure 2. Their molecular weights are listed in Table 1. Concentrated (0.10 wt.%) stock solutions were prepared no more than 7 days prior to use: Powder flocculant (0.2 g) was added slowly into the vortex of deionized water (200 mL), with the stirring speed at 600 rpm for 5 min. Afterwards, the stirring speed was reduced to 250 rpm and maintained for another 12 h. The stock solutions were sealed in a glass beaker using parafilm at room temperature before use. The dosage of flocculants added into the quartz slurry is expressed as g/t (grams of flocculant per ton of quartz solids), and it varies from 10 to 100 g/t.

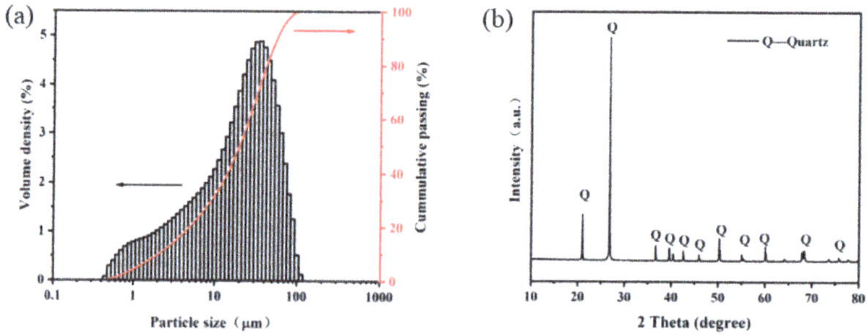

Figure 1. (**a**) Particle Size Distribution (PSD); (**b**) XRD pattern of quartz sample.

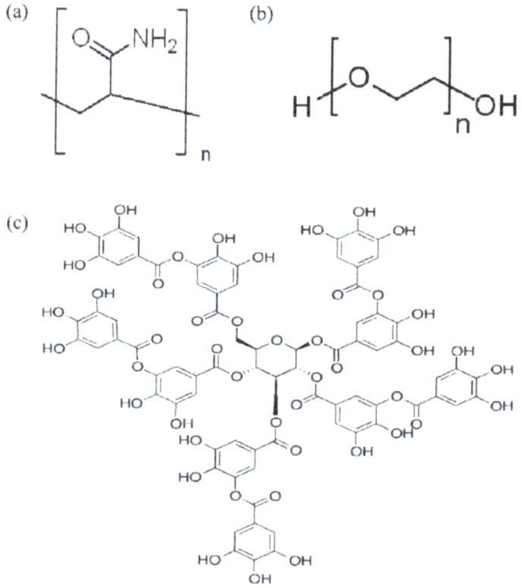

Figure 2. Molecule structure: (**a**) PAM; (**b**) PEO; (**c**) TA.

Table 1. Flocculants/Cofactors used in this study.

	Molecular Weight, g/mol	Type
PEO	7 million	Powder
PAM	10~12 million	Powder
MMT	low	Powder
TA	1701.20	Powder

Montmorillonite (MMT) is a nanolayered phyllosilicate mineral that consists of two O-Si-O tetrahedral sheets sandwiching one O-Al(Mg)-O octahedral sheet, as shown in Figure 3. Isomorphic substitutions always exist. For example, Si^{4+} in the tetrahedral sheet can be replaced by Al^{3+}; Al^{3+} in the octahedral sheet can be replaced by Mg^{2+} [17]. Due to such an isomorphous substitution, the MMT nanolayer exhibits inherent negative charges, which can be compensated for by cations (Na^{+}, Ca^{2+}) in the interlayer space [18,19]. It is noted that these negative charges are permanent, differing from carboxylic acid which cannot be dissociated below its pKa. In particular, the van der Waals force and electrostatic force holding the MMT nanolayers together can be overcome by certain methods. For example, MMT particles disassembled in aqueous dispersion, leading to reduced particle size. Ultimately, the fine MMT particles can be exfoliated to an individual nanolayer [20]. Figure 4a shows the PSD of the MMT sample. It was found that the water dispersed sample had a D_{90} of 4.521 μm and D_{50} of 0.842 μm, indicating the presence of single or multiple nanolayers. Larger particles may be attributed to MMT aggregates or the presence of impurities, including quartz and albite. As shown in the XRD pattern in Figure 4b, the MMT sample had a $d_{(001)}$-spacing of 14.579 Å ($2\theta = 6.057°$), which is a mixture of Ca^{2+} -MMT and Na^+ -MMT [21]. The MMT slurry was stirred for at least 12 h prior to use for ample hydration and exfoliation.

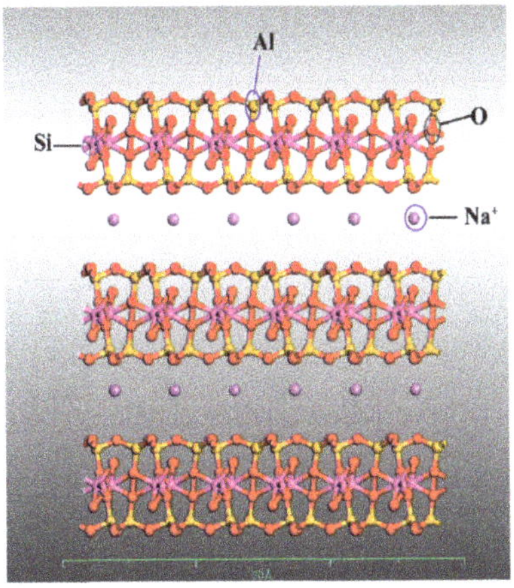

Figure 3. Crystal structure of montmorillonite.

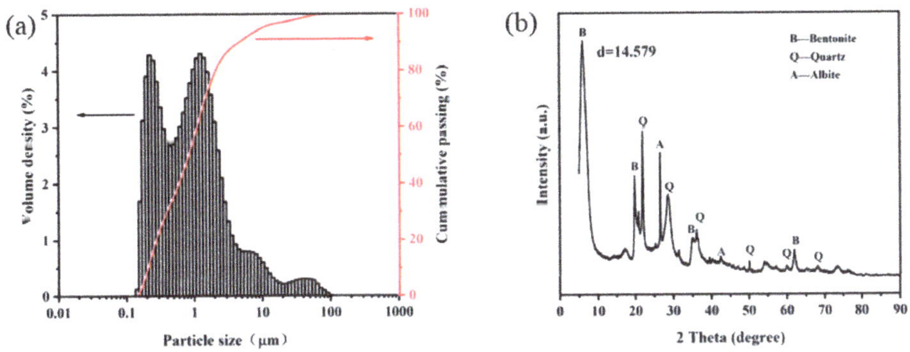

Figure 4. (**a**) Particle size distribution; (**b**) XRD pattern of montmorillonite sample.

2.2. Flocculation Sedimentation Tests

Slurries were prepared by adding a known amount of the quartz sample to 100 mL of water in a 150 mL glass beaker (diameter 5.8 cm, height 8.8 cm) fitted with a C50 stir bar (diameter 8 mm, length 5 cm). The pH of the slurry was adjusted to 1.5 using sulfuric acid, then it was preconditioned at a high mixing intensity (600 rpm) for 15 min. In this work, 10 g of the closed-circuit ground 200-mesh quartz sample was added to make a 10 wt.% slurry, while 1 g of the 2000-mesh quartz sample was added to make a 1 wt.% slurry.

After the high intensity mixing, the stirring speed was decreased to 200 rpm prior to the addition of the flocculants, then stirring was continued for 1 min. Where applicable, cofactors were added 1 min prior to the addition of flocculants at a constant stirring rate. The flocculated slurry was then poured into a 100 mL measuring cylinder and settled naturally. Settling was allowed to proceed for various selected durations, then 2 mL of supernatant was carefully syringed 1 cm underneath the water surface. The sampled supernatant was diluted where necessary for the turbidity measurement (LEI-CI WZB-175).

2.3. FBRM Measurement

The particle and aggregate size at different stages of flocculation were examined in situ using the FBRM, the principle of which has been described previously by Fawell, P.D. et al. [22,23]. All FBRM measurements were carried out using a Mettler Toledo G400 field unit fitted with a laboratory probe. The probe has a flat sapphire window of 9.5 mm diameter located at the tip of the stainless-steel body (136/70 mm long, 14/9.5 mm diameter).

A polytetrafluoroethylene container (internal diameter 7.0 cm, height 7 cm) was custom built to accommodate the FBRM probe in a horizontal position near the inner wall of the container, 1 cm above the bottom and away from the C50 stir bar. The probe had a 45° angle to the tangent direction of the flow speed to avoid flocs adhering to the sapphire window, as shown in Figure 5. Such configuration of the probe tip not only ensures good sample presentation but also allows slow-stirred slurry to be measured. This is due to excessively large flocs settling down to the bottom of the container easily during slow stirring; however, obvious breakage of flocs occurs once sped up.

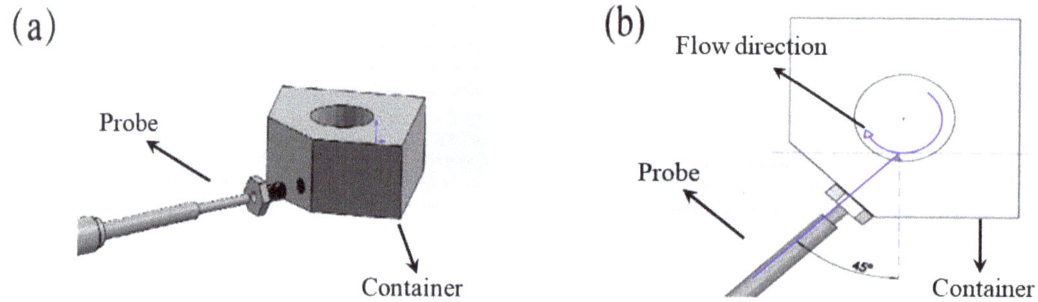

Figure 5. Configuration of custom-built container to accommodate the FBRM probe: (**a**) stereogram; (**b**) vertical view.

The FBRM data was processed and analyzed using Version 4.4 acquisition software. It was continuously scanned with a 2 s duration over the full range of chord lengths (1 to 4000 μm). Chord Lengths Distributions are expressed as line graphs for simplicity. Both unweighted and square-weighted chord lengths were acquired using the software. The unweighted chord length distribution is number-sensitive, providing an implication for the fine particles that are suspended in the slurry. Square-weighted chord length distribution represents the volumetric and mass contribution of the large flocs in the slurry.

Primary mode and macro mode measure flocs in a different way. The primary mode is suitable for detecting the suspended fine particles, while the response of macro mode is too slow to capture fine particles. With respect to the primary mode, each section of the porous large floc is considered as a single chord length, while the slower macro response tends to detect each floc as a single particle with a single chord length. When we are aiming to monitor the growth and breakage of flocs (flocculation kinetics), the macro mode with volumetric weightings is recommended. Alternatively, unweighted chord length distributions in primary mode are preferred if we need to measure the counts of fine particles [24].

2.4. Zeta Potential and Effective Diameter Measurements

The zeta potential of quartz dispersions was measured using a NanoPlus3 (Micromeritics Instrument, Norcross, GA, USA). The quartz dispersion was prepared by adding 5000-mesh fine particles to 10^{-3} mol/L KNO_3 solution at a solid concentration of 0.5 mg/mL, sonicating for 30 min, and allowing it to stand for 24 h. Afterwards, the supernatant was transferred into a disposable cell for zeta potential measurement. The cell was kept at 25 °C and equilibrated for 120 s prior to measurement. The pH of dispersion was regulated by adding a suitable amount of 10^{-1} M NaOH or HNO_3; a desired amount of MMT was added, followed by conditioning for 30 min.

The effective diameters of the flocculant's molecule and their associated complexes in aqueous solution were determined using a Malvern ZS90. The effective diameter is the hydrodynamic diameter of a polymer molecule conformation. For a polydisperse system, the effective diameter measured is a value calculated from the averaged intensity of the scattered light for each particle. The 0.1 wt.% PEO/PAM stock solutions were prepared, as described above, 12 h before being added into the cuvette. The 0.1 wt.% TA was mixed with PEO stock solution and stirred at 200 rpm for 1 min before measurement.

3. Results and Discussion

3.1. Flocculation Sedimentation

The settling behaviors of quartz slurry were investigated to examine the flocculation efficiency of the proposed polymers–cofactors system. The supernatant turbidity of 10 wt.% -200-mesh quartz slurry during sedimentation is shown in Figure 6. These sedimentation curves can be described using the exponential decay of turbidity as a function of time. The use of PEO at a dosage of 10~100 g/t settles significantly faster, with 10-min turbidity reduced to 94.4 NTU (100 g/t), compared to 834 NTU (100 g/t) at the same dose of PAM. Low turbidity indicated a better clarity of the supernatant, containing a smaller number of suspended particles. In the simulated acidic leaching pulp, PEO was proved to be more efficient to flocculate quartz particles. On the other hand, PAM might be good at accelerating settling rate, but it was not capable of controlling the turbidity of the supernatant, even with an excessive dosage.

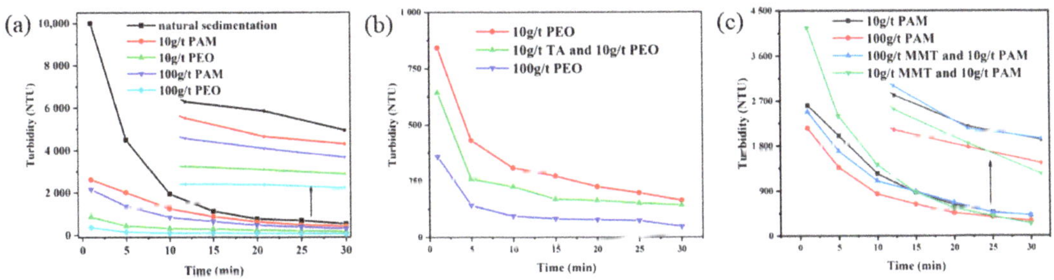

Figure 6. Supernatant turbidity of 10 wt.% -200-mesh quartz slurry during a 30-min sedimentation: (**a**) high molecule flocculants; (**b**) PEO and cofactor; (**c**) PAM and cofactor.

It is seen from Figure 6b that the combination use of TA-PEO gave a clearer supernatant with further decreased turbidity compared to that of PEO alone. It suggested that TA could promote the formation of larger quartz flocs using PEO, causing faster settling rates and entrapping more fine suspended particles. Such a synergistic effect was, however, not found for the combination use of MMT-PAM. The turbidity was even higher after the addition of MMT into the PAM system, as seen from Figure 6c. These extra suspended fine particles were probably contributed by the slow settling MMT nanolayers.

The settling behaviors of 1.0 wt.% 2000-mesh quartz slurry were also studied to examine the flocculation efficiency of fine particles, mimicking the clarification of turbid thickener overflow. The results in Figure 7a show that PEO was still very effective in flocculating fine particles. A lower supernatant turbidity can be obtained with only 10 g/t PEO than its PAM counterpart at the same dosage. The effect of TA addition to PEO was identical to that of a -200-mesh quartz system, significantly decreasing the turbidity. For example, the addition of 10 g/t TA to 10 g/t PEO exhibited lower turbidity (470 NTU at 5-min sedimentation). It is noted that settling accelerated after the first several minutes for PEO and PEO-TA systems, indicating a longer incubation time for floc formation than other flocculants. It is seen from Figure 7c that adding any amount of MMT did not accelerate the settling of fine quartz particles based on 10 g/t PAM. On the other hand, the addition of 10 g/t MMT prior to 100 g/t PAM dramatically accelerated settling with much lower turbidity of the supernatant (402 NTU at 10-min sedimentation). The possibility of failure in the PAM-MMT in 200-mesh quartz slurry because most of the MMT was consumed by coarse particles which settled easily.

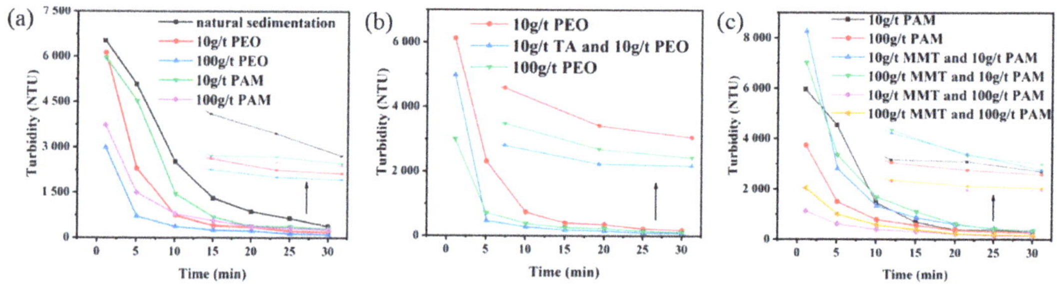

Figure 7. Supernatant turbidity of 1 wt.% -2000-mesh quartz slurry during a 30-min sedimentation: (**a**) high molecule flocculants; (**b**) PEO and cofactor; (**c**) PAM and cofactor.

3.2. Flocs Formation and Breakage

Figure 8 shows the mean particle size (MPS) and particle size distribution (square weighted, macro mode) evolution as a function of time for 10 wt.% 200-mesh quartz slurry at 400 rpm. As can be seen, a dramatic increase of MPS was observed upon the addition of flocculants (100 g/t PEO at 90 s). During the flocculation stage (90~106 s), the flocs rapidly grew to reach a maximum size. The maximum MPS was about 230 μm under the tested condition. The floc size soon experienced a continuous decrease upon exposure to a longer reaction time at a constant shear rate. At the end of the measured period (300 s), the mean floc size decreased to ~118 μm, not far from its original value without any addition of flocculants.

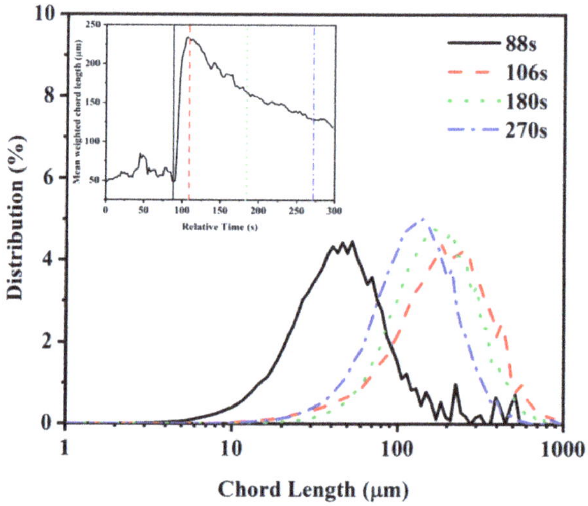

Figure 8. The trend of Mean Particle Size and PSD (square weighted, macro mode) as a function of stirring time: 100 g/t PEO, 400 rpm.

Figure 9 demonstrates the FBRM results of different flocculants/cofactors systems that used quartz flocculation at a stirring speed of 200 rpm. The cofactors (TA/MMT) were added 60 s prior to flocculant (PEO/PAM) additions, and all reagents were used at a dosage of 100 g/t. As can be seen, the particle size did not change in the period from 0 s to 90 s, indicating that cofactors alone were incapable of flocculating quartz particles without the presence of high molecule weight flocculants.

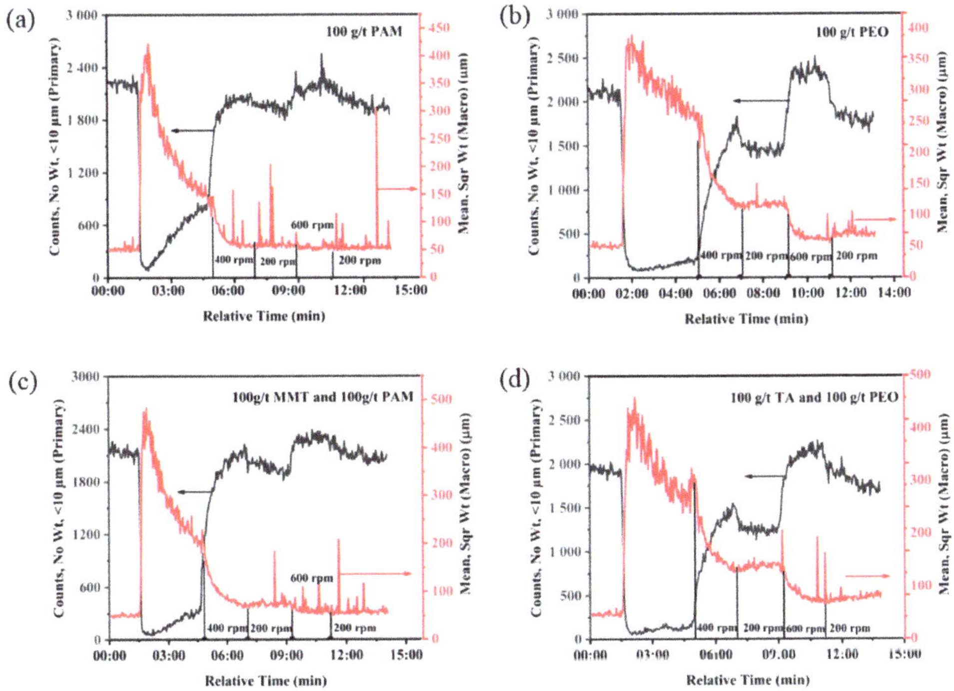

Figure 9. The trend of Mean Particle Size (MPS) (square weighted, macro mode) and Counts of Fine Particles (CFP) (≤10 µm, no weighted, primary mode) as a function of stirring time: initially 200 rpm and accelerated afterwards: (**a**) PAM; (**b**) PEO; (**c**) PAM-MMT; (**d**) PEO-TA.

The maximum floc size reached ~410 µm by adding PAM, which was slightly larger than that formed by PEO. Upon the addition of MMT, the MPS increased sharply to more than 450 µm, enhancing the flocculation of quartz particles by PAM. Additionally, the addition of TA leads to larger floc size as a result of the synergistic effect between TA and PEO. The floc size in all situations then experienced a continuous decrease upon exposure to a longer reaction time, and the slope of the decrease was calculated and listed in Table 2. It was found that PEO and PEO-TA systems exhibit a much smaller slope (absolute value) compared to their PAM and PAM-MMT counterparts. It means that PEO and PEO-TA systems experienced a weaker degree of aggregate rupture during longer reaction times after the peak magnitude.

Table 2. Slopes of MPS decrease and CFP increase.

		PAM	PEO	PAM-MMT	PEO-TA
200 rpm	MPS	−2.08	−0.76	−2.25	−1.27
	CFP	4.92	0.45	1.42	0.49
400 rpm	MPS	−1.98	−0.87	−2.09	−0.59
	CFP	13.66	7.89	14.13	7.31

Counts of Fine Particles (CFP) ($\leq$10 μm, no weighted, primary mode) were calculated using FBRM software and the results for quartz flocculation at 200 rpm were also presented in Figure 9. Particles finer than 10 μm contributed to most of the turbidity after settling and their counts sharply decreased to a very low level upon the addition of any type of flocculant (from more than 2000 to less than 200). However, the CFP soon increased to much higher values in the case of PAM, as shown in Figure 9a. The addition of MMT can be helpful to keep the low number of fine particles for PAM with only a small increase after 2~3 min of shear action. During conditioning with PEO and PEO-TA at 200 rpm, the CFP were almost constant at a very low level, reflecting the stability of flocs. For comparison, the slope of the CFP increase was also calculated and listed in Table 2.

The results of the FBRM corresponded well with flocculation behaviors based on turbidity measurement. It indicated that PEO-TA and PAM-MMT acted synergistically to create more stable flocs from the quartz suspension with high resistance to floc breakage when subject to shear force. It is seen from Figure 9 that the flocs formed quickly but broke quickly under the use of PAM alone. In practice, the mixing reaction time of slurry and PAM in the feed pipe and feed well can last longer than 1 min, resulting in significantly broken flocs. The PEO and its combination with TA, as well as the PAM-MMT combination, help the flocculation of quartz slurry due to their higher shear resistance.

In addition to the molecule type and their combination, the flocculation process can also be affected by hydrodynamics, especially turbulence intensity. Figure 9 also shows full profiles for quartz flocculation switching to a higher stirring rate after the initial 200 rpm. Generally, higher shear rates (400 rpm and 600 rpm) can give a faster rate of floc breakage, together with the sudden release of fine particles. Such degradation of flocs at high turbulence occurs following such an order: PAM > PAM-MMT > PEO > PEO-TA.

When the lower shear rate (200 rpm) was restored, the floc size remained unchanged. This indicated that nonionic PAM/PEO-induced flocculation was irreversible upon high shear, and the flocs did not regrow after breakage. After resuming the initial lower shear rate, the CFP slightly reduced. Reflocculation may occur but not in the form of floc size regrow.

This was consistent with expectations that high-weight-molecule-flocculating solid particles can be effectively deactivated as a result of breakage and unable to bridge these particles again. These results shed light on the critical differences between coagulation and flocculation [25]. Aggregates are produced as a result of charge neutralization for coagulation, and a reduction in shear intensity usually leads to the regrowth of aggregate while a new equilibrium can be restablished. In contrast, flocs formed by polymer-bridging does not lead to a new equilibrium once the bridging ruptures, and the dismantlement of fine particles is irreversible without any additional flocculant dosage.

Applying a cycled shear scheme as shown in Figure 9 is a good tactic to study the floc strength and floc reversibility under different turbulent conditions [26,27]. The above results indicate that only fine flocs can be partially reflocculated after breakage. In order to calculate the MPS decrease slope and CFP increase slope at higher shear strength, FBRM test work was conducted for freshly prepared quartz slurry at a constant stirring speed of 400 rpm, and the results are shown in Figure 10. The corresponding slopes in Table 2 further indicate that PEO and its combination with TA outperformed PAM and PAM-MMT, in terms of floc strength. However, it is better to avoid either intense turbulent condition or long reaction time. Judicious feed well optimization allows aggregate growth to take place under the milder shear experienced immediately below the feed well exit. The implementation of the FBRM online particle size measurement allows design options or flocculant addition points to be identified that produce the optimal residence time and flocculation response [28,29].

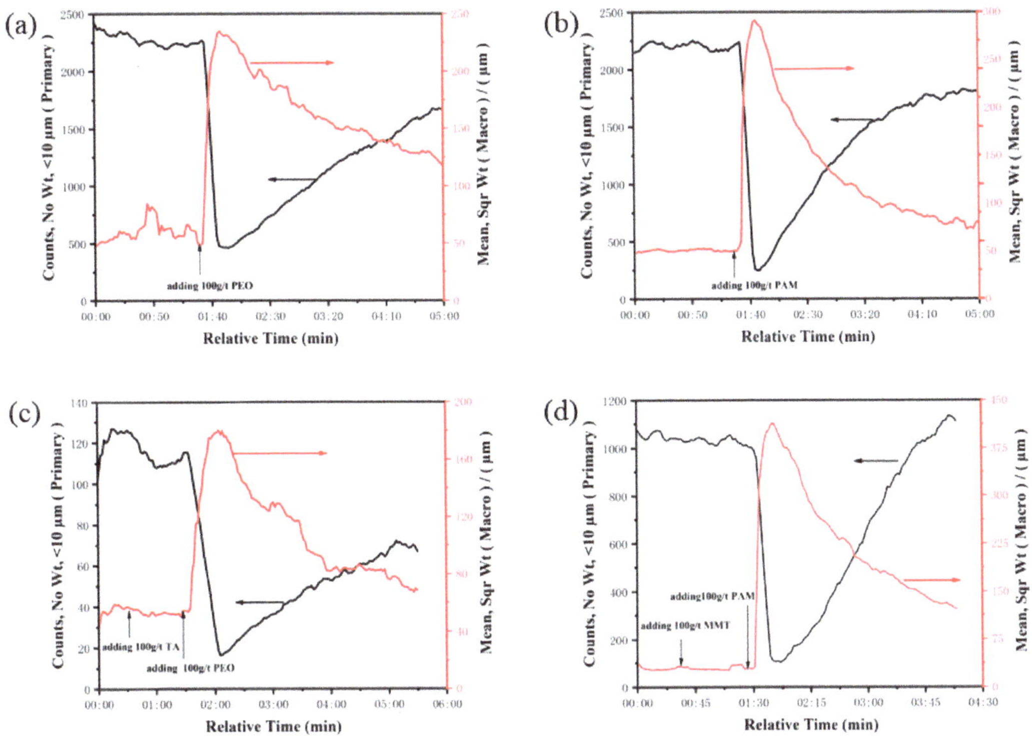

Figure 10. The trend of Mean Particle Size (square weighted, macro mode) and Counts of Fine Particles (≤10 µm, no weighted, primary mode) as a function of stirring time: 400 rpm: (**a**) PEO; (**b**) PAM; (**c**) PEO-TA; (**d**) PAM-MMT.

3.3. Mechanisms of Flocculation

To achieve the synergistic function of the flocculants/cofactors dual reagents system in quartz flocculation, several pathways were possible. One way is that cofactors adsorb onto the quartz surface acting as the bridge and facilitating the adsorption of flocculants. Another possible pathway is that cofactors act as coagulant, neutralizing surface charge.

Zeta potential is an important parameter of the stability of colloidal suspensions. Figure 11 shows the zeta potential of quartz at pH 1.5 and pH 7, respectively. Quartz particles were negatively charged at pH 7 (−36.95 mV), but positively charged at pH 1.5 (11.04 mV). It has been reported that the point of zero charge (PZC) for quartz suspension is pH 2, below which silanol groups (-Si-O$^-$) are dominant, rendering its surface negatively charged, while these silanol groups can be protonated at acidic pH, resulting in a positively charged surface [11,26]. It is generally accepted that the stability of colloidal dispersion depends on surface charge, with zeta potential of ≤±10 mV corresponding to unstable dispersion, and ≥±30 mV corresponding to a stable dispersion as a result of the electrostatic repulsion [6]. The present quartz suspension should be relatively unstable without flocculants; however, it naturally settles slowly, which is still inadequate for high throughput thickening performance.

A reduction in the magnitude of the zeta potential was observed with the increasing MMT concentration in acidic conditions (2.74 mV for 10 g/t and −0.63 mV for 100 g/t). The absence of electrostatic repulsion would cause the system to aggregate under van der Waals attractive forces. The quartz suspension can then be destabilized to some extent by MMT adsorption, due to a reduction in the magnitude of the surface charge.

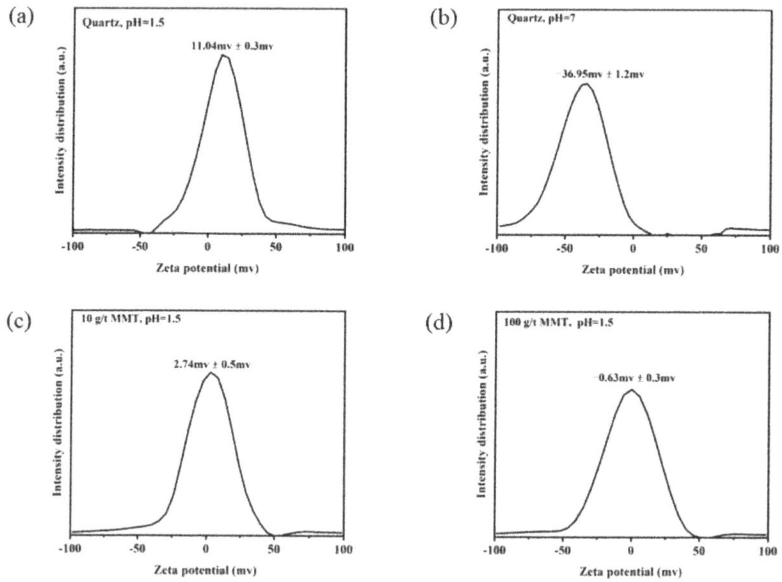

Figure 11. The zeta potential of quartz suspension: (**a**) pH1.5; (**b**) pH7; (**c**) pH1.5 with 10g/t MMT; (**d**) pH1.5 with 100g/t MMT.

The addition of MMT significantly improved the flocculation efficiency by PAM. However, the authors believed that coagulation caused by MMT's charge neutralization might not be the main contributor for an improvement. Ionic sulfonated salts, SDBS or AMPS, can also neutralize surface positive charge in acid media; however, a very large amount of these sulfonated salts (more than 1000 g/t) has to be used and the flocculation performance is still not as good as 10 g/t MMT in the PAM system (results not shown here). Considering the magnitude of the zeta potentials (11.04 mV) of the quartz suspension was not a characteristic of a highly stable system; the surface modification upon MMT adsorption and its interaction with PAM could be the dominant contribution for a better flocculation efficiency.

The PAM molecule was nonionic and linear, and bridging flocculation by PAM for the quartz suspension was generally driven via hydrogen bonds formed between the PAM's amide oxygen and quartz surface silanol groups. Due to the low binding energy, they could not stand the high shear rates such that PAM molecules detached from quartz particles. Once broken, these bonds may not be able to be reformed, so the broken quartz flocs could no longer be held together. Compared to the adsorption energy of PAM on the (101) face of quartz crystalline (−16.93 kJ/mol), the adsorption energy of PAM on the (001) face of MMT crystalline (O-Si-O tetrahedral sheets) is −49.16 kJ/mol [30]. Therefore, the interaction of MMT-modified quartz surface with PAM was greatly enhanced. Sedimentation tests carried out for the 1.0 wt.% 2000-mesh MMT suspension using PAM as flocculants is shown in Figure 12. It can be seen that the MMT suspension was easily flocculated with a much faster settling rate and much lower supernatant turbidity, in comparison with quartz under the same condition. Therefore, the improved flocculation behavior of the MMT addition to the PAM system was due to the stronger hydrogen binding between MMT and PAM than that of quartz and PAM, together with electrostatic attraction between MMT and quartz. Adding SDBS/AMPS can also lead to their adsorption on a quartz surface; however, the van der Waals force existing between these organic small molecules and PAM was much weaker than MMT's hydrogen binding.

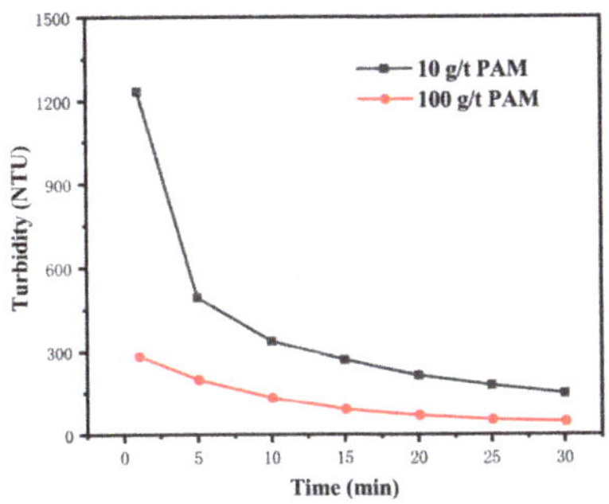

Figure 12. Supernatant turbidity of 1 wt.% -2000-mesh montmorillonite slurry during a 30-min sedimentation.

The synergistic function of the PEO-TA system can be different from that of the PAM-MMT system. The PEO itself behaved better than the PAM because of the stronger hydrogen binding of PEO's ether oxygen with the silanol-bearing surface [31–33]. The PAM has a higher molecule weight than the PEO, possibly associated with higher viscosity. In this work, adsorption is a more determining factor, compared to molecule weight, in the process of fine silica particle flocculation. The TA had a negligible affinity for quartz, and in such circumstances, previous research has proven that TA barely adsorbed on quartz [27]. It was proposed in the literature that TA interacted with PEO to form associative polymer complexes which then adsorbed onto a mineral surface through PEO [15,16]. Polymeric flocculant dissolved in aqueous solution has a measurable particle size due to molecule structure conformation. Figure 13 shows the molecule size of flocculants measured by dynamic light scattering. It is seen that the formed PEO-TA complexes were larger in size compared to PEO alone, and such a complex structure was more efficient to increase the floc size. In addition, the PEO-TA complexes were comparable with PAM in size, indicating that the formed complexes might have a molecule weight of around 10~12 millions.

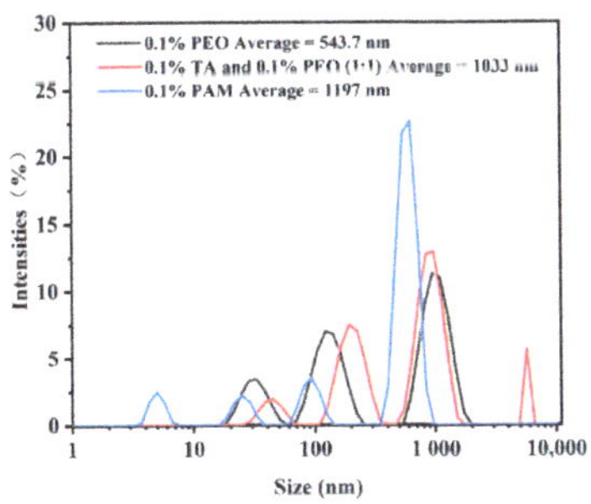

Figure 13. Dynamic Light Scattering measurement: PEO, TA-PEO, and PAM.

The bridging and sweeping between quartz particles can be further promoted by a higher molecule weight flocculant, thus promoting flocculation. A similar conclusion was made in the previous publication for a quartz suspension, particularly in neutral and alkaline suspensions [27]. The present work expanded the application of PEO-TA complexes to more acidic media that could be helpful for acid leaching pulp. It was also proposed that phenolic compounds can increase the stiffness of PEO molecule chains and make them adsorb more easily onto fine particles [16]. Regarding the structure of the TA-PEO associative complex, the initially flexible linear PEO chains may become rigid due to their association with TA. Complexes consisting of well-dissolved PEO entanglements clustered by cofactors ensure the large size of the complexes. As can be seen from Figure 11, quartz has a low charge density; therefore, its affinity is strong for neutral polymers such as PEO, resulting in adsorption with more trains and loops [34]. The result was that these complex molecules were bonded more firmly with quartz particles (or less likely to desorb), developing large flocs that were more resistant against shear force. The mechanisms underlying the synergistic function of a flocculants/cofactors dual reagents system in quartz flocculation were schematically illustrated in Figure 14.

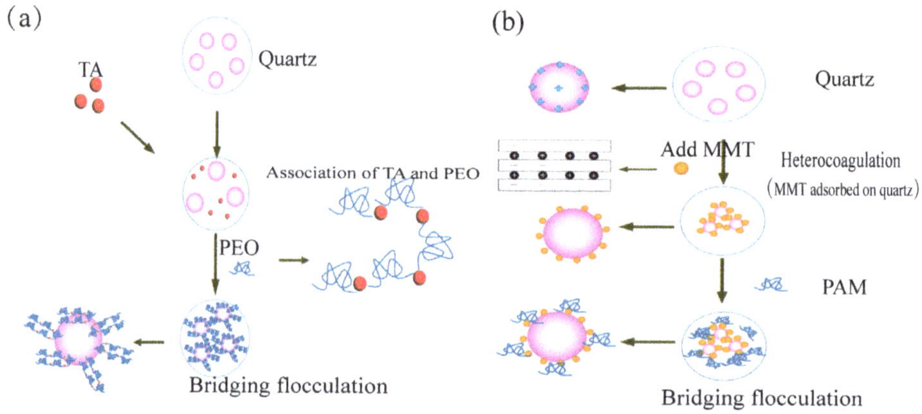

Figure 14. Schematic illustration of flocculation: (**a**) Quartz-TA-PEO; (**b**) Quartz-MMT-PAM.

4. Conclusions

The flocculation behaviors of quartz in simulated acid leaching pulp using nonionic flocculant poly(ethylene oxide) (PEO) with a combination of tannic acid (TA) as the cofactor and another combination of nonionic polyacrylamide (PAM) and a negatively charged montmorillonite (MMT) nanosheet were investigated in this study, with dynamic floc size monitoring using the in situ focused beam reflectance measurement technique.

In terms of polymeric flocculants, PEO outperforms PAM by lowering the turbidity of the suspension after flocculation sedimentation, because of the stronger affinity to quartz by the hydrogen bond between the ether oxygen and silanol group. The addition of TA significantly improves the flocculation efficiency of quartz by PEO, and MMT improves the flocculation efficiency by PAM. Flocs breakage upon turbulent conditions plays an important role in the flocculation performance. The higher shear resistance provided by PEO/PEO + TA/PAM + MMT leads to a more compact floc structure with less reduction in floc size and less release of fines upon shear force and, subsequently, clearer suspension upon sedimentation.

However, mechanisms underlying the increase of floc strength by cofactors between TA and MMT can be different in the combinations of PEO + TA and PAM + MMT, respectively. Dynamic light scattering and adsorption measurements confirm that the pathway of the PEO + TA flocculation is via the initial formation of associative complexes between tannic acid and PEO in a solution with the formation of entangled and large-sized clusters, and then bridging of quartz particles by the associative complexes more effectively. The zeta potential measurement and molecule simulation indicate that the pathway of the PAM + MMT flocculation is via MMT nanosheet adsorption on quartz by electrostatic forces, and then bridging of the surface-modified quartz particles more effectively by PAM, due to the more negative adsorption energy of the acrylamino group on tetrahedral sheets. Coagulation might not play a dominant role in PAM + MMT flocculation; although, neutralization of positively charged quartz by a negatively charged MMT nanosheet occurs to some extent. Furthermore, reformation of broken flocs was difficult for both flocculation systems.

Author Contributions: Conceptualization, B.G. and K.J.; methodology, X.Z. and B.G.; software, X.Z.; validation, H.X., R.D. and X.Z.; formal analysis, B.G.; investigation, B.G. and K.J.; resources, B.G. and K.J.; data curation, X.Z.; writing—original draft preparation, X.Z.; writing—review and editing, X.Z. and B.G.; visualization, X.Z. and B.G.; supervision, B.G. and K.J.; project administration, B.G. and H.X.; funding acquisition, H.X. and B.G. All authors have read and agreed to the published version of the manuscript.

Funding: This work is supported by the National Key Research and Development Program (No. 2022YFC2904604) from the Ministry of Science and Technology of China.

Data Availability Statement: Data are available upon reasonable, by the corresponding authors.

Acknowledgments: The SNF was kindly acknowledged for providing flocculants samples.

Conflicts of Interest: The authors declare no conflict of interest.

References

1. Havlik, T. *Hydrometallurgy: Principles and Applications*; Woodward Publishing: Cambridge, UK, 2014.
2. Kazadi, D.M.; Groot, D.R.; Steenkamp, J.D.; Pöllmann, H. Control of silica polymerisation during ferromanganese slag sulphuric acid digestion and water leaching. *Hydrometallurgy* **2016**, *166*, 214–221. [CrossRef]
3. Ritcey, G.M. Crud in solvent extraction processing—A review of causes and treatment. *Hydrometallurgy* **1980**, *5*, 97–107. [CrossRef]
4. Fletcher, A.W.; Gage, R.C. Dealing with a siliceous crud problem in solvent extraction. *Hydrometallurgy* **1985**, *15*, 5–9. [CrossRef]
5. Kitchener, J.A. Principles of action of polymeric flocculants. *Br. Polym. J.* **1972**, *4*, 217–229. [CrossRef]
6. Lee, C.S.; Robinson, J.; Chong, M.F. A review on application of flocculants in wastewater treatment. *Process Saf. Environ. Prot.* **2014**, *92*, 489–508. [CrossRef]
7. Hogg, R. Flocculation and dewatering. *Int. J. Miner. Process.* **2000**, *58*, 223–236. [CrossRef]
8. Moud, A.A. Polymer based flocculants: Review of water purification applications. *J. Water Process Eng.* **2022**, *48*, 102938. [CrossRef]
9. Loan, M.; Newman, O.M.G.; Cooper, R.M.G.; Farrow, J.B.; Parkinson, G.M. Defining the Paragoethite process for iron removal in zinc hydrometallurgy. *Hydrometallurgy* **2006**, *81*, 104–129. [CrossRef]
10. Dyer, L.G.; Richmond, W.R.; Fawell, P.D. Simulation of iron oxide/silica precipitation in the paragoethite process for the removal of iron from acidic zinc leach solutions. *Hydrometallurgy* **2012**, *119–120*, 47–54. [CrossRef]
11. Valenzuela-Elgueta, J.; Delgado, A.V.; Ahualli, S. Effect of cationic surfactant addition on the electrokinetics and stability of silica/kaolinite suspensions in copper hydrometallurgy conditions. *Miner. Eng.* **2021**, *169*, 106958. [CrossRef]
12. Figdore, P.E. Adsorption of surfactants on kaolinite: NaCl versus $CaCl_2$ salt effects. *J. Colloid Interface Sci.* **1982**, *87*, 500–517. [CrossRef]
13. Besra, L.; Sengupta, D.K.; Roy, S.K.; Ay, P. Flocculation and dewatering of kaolin suspensions in the presence of polyacrylamide and surfactants. *Int. J. Miner. Process.* **2002**, *66*, 203–232. [CrossRef]
14. Cong, R.; Pelton, R. The influence of PEO/poly(vinyl phenol-co-styrene sulfonate) aqueous complex structure on flocculation. *J. Colloid Interface Sci.* **2003**, *261*, 65–73. [CrossRef] [PubMed]
15. Gaudreault, R.; van de Ven, T.G.M.; Whitehead, M.A. Mechanisms of flocculation with poly(ethylene oxide) and novel cofactors. *Colloids Surf. A Physicochem. Eng. Asp.* **2005**, *268*, 131–146. [CrossRef]
16. van de Ven, T.G.M. Association-induced polymer bridging by poly(ethylene oxide)–cofactor flocculation systems. *Adv. Colloid Interface Sci.* **2005**, *114–115*, 147–157. [CrossRef] [PubMed]
17. Zhou, C.; Tong, D.; Yu, W. 7-Smectite Nanomaterials: Preparation, Properties, and Functional Applications. In *Nanomaterials from Clay Minerals*; Wang, A., Wang, W., Eds.; Elsevier: Amsterdam, The Netherlands, 2019; pp. 335–364.
18. Peng, C.; Min, F.; Liu, L. Effect of pH on the adsorption of dodecylamine on montmorillonite: Insights from experiments and molecular dynamics simulations. *Appl. Surf. Sci.* **2017**, *425*, 996–1005. [CrossRef]
19. Emmerich, K.; Koeniger, F.; Kaden, H.; Thissen, P. Microscopic structure and properties of discrete water layer in Na-exchanged montmorillonite. *J. Colloid Interface Sci.* **2015**, *448*, 24–31. [CrossRef]
20. Shi, D.; Yu, W.; Li, R.K.Y.; Ke, Z.; Yin, J. An investigation on the dispersion of montmorillonite (MMT) primary particles in PP matrix. *Eur. Polym. J.* **2007**, *43*, 3250–3257. [CrossRef]
21. Oueslati, W.; Ben Rhaiem, H.; Lanson, B.; Ben Haj Amara, A. Selectivity of Na–montmorillonite in relation with the concentration of bivalent cation (Cu^{2+}, Ca^{2+}, Ni^{2+}) by quantitative analysis of XRD patterns. *Appl. Clay Sci.* **2009**, *43*, 224–227. [CrossRef]
22. Heath, A.; Fawell, P.; And, P.B.; Swift, J. Estimating Average Particle Size by Focused Beam Reflectance Measurement (FBRM). *Part. Part. Syst. Charact.* **2002**, *19*, 84–95. [CrossRef]
23. Grabsch, A.F.; Yahyaei, M.; Fawell, P.D. Number-sensitive particle size measurements for monitoring flocculation responses to different grinding conditions. *Miner. Eng.* **2020**, *145*, 106088. [CrossRef]
24. Kyoda, Y.; Costine, A.D.; Fawell, P.D.; Bellwood, J.; Das, G.K. Using focused beam reflectance measurement (FBRM) to monitor aggregate structures formed in flocculated clay suspensions. *Miner. Eng.* **2019**, *138*, 148–160. [CrossRef]

25. Spicer, P.T.; Pratsinis, S.E.; Raper, J.; Amal, R.; Bushell, G.; Meesters, G. Effect of shear schedule on particle size, density, and structure during flocculation in stirred tanks. *Powder Technol.* **1998**, *97*, 26–34. [CrossRef]
26. Wang, C.; Sun, C.; Liu, Q. Formation, breakage, and re-growth of quartz flocs generated by non-ionic high molecular weight polyacrylamide. *Miner. Eng.* **2020**, *157*, 106546. [CrossRef]
27. Wang, D.; Wang, D.; Deng, C.; Wang, K.; Tan, X.; Liu, Q. Flocculation of quartz by a dual polymer system containing tannic acid and poly(ethylene oxide): Effect of polymer chemistry and hydrodynamic conditions. *Chem. Eng. J.* **2022**, *446*, 137403. [CrossRef]
28. Owen, A.T.; Fawell, P.D.; Swift, J.D.; Labbett, D.M.; Benn, F.A.; Farrow, J.B. Using turbulent pipe flow to study the factors affecting polymer-bridging flocculation of mineral systems. *Int. J. Miner. Process.* **2008**, *87*, 90–99. [CrossRef]
29. Swift, J.D.; Simic, K.; Johnston, R.R.M.; Fawell, P.D.; Farrow, J.B. A study of the polymer flocculation reaction in a linear pipe with a focused beam reflectance measurement probe. *Int. J. Miner. Process.* **2004**, *73*, 103–118. [CrossRef]
30. Huang, Y. *Study on Settlement Mechanism of Fine Kaolinite Tailing Ballasteded Flocculation by Quartz Carrier*; China University of Mining and Technology: Beijing, China, 2019.
31. Mpofu, P.; Addai-Mensah, J.; Ralston, J. Flocculation and dewatering behaviour of smectite dispersions: Effect of polymer structure type. *Miner. Eng.* **2004**, *17*, 411–423. [CrossRef]
32. Mpofu, P.; Addai-Mensah, J.; Ralston, J. Investigation of the effect of polymer structure type on flocculation, rheology and dewatering behaviour of kaolinite dispersions. *Int. J. Miner. Process.* **2003**, *71*, 247–268. [CrossRef]
33. Rubio, J.; Kitchener, J.A. The mechanism of adsorption of poly(ethylene oxide) flocculant on silica. *J. Colloid Interface Sci.* **1976**, *57*, 132–142. [CrossRef]
34. Quezada, G.R.; Piceros, E.; Saavedra, J.H.; Robles, P.; Jeldres, R.I. Polymer affinity with quartz (101) surface in saline solutions: A molecular dynamics study. *Miner. Eng.* **2022**, *186*, 107750. [CrossRef]

minerals

Article

Microflotation of Fine Rutile and Garnet with Different Particle Size Fractions

Zheyi Zhang [1,2], Liuyi Ren [1,3,*], Yimin Zhang [1,3] and Shenxu Bao [1,3]

1 School of Resources and Environmental Engineering, Wuhan University of Technology, Wuhan 430070, China
2 School of Minerals Processing and Bioengineering, Central South University, Changsha 410083, China
3 Hubei Key Laboratory of Mineral Resources Processing and Environment, Wuhan 430070, China
* Correspondence: rly1015@whut.edu.cn or rly1015@163.com

Abstract: In this paper, −10 μm rutile and −30 μm garnet particles were selected as samples. The effects of different reagents on the flotation of rutile and garnet single minerals were studied, and the mechanism was analyzed by the contact angle, zeta potential, and Fourier transform infrared (FTIR) measurements. The flotation results show that an optimal recovery is obtained with benzohydroxamic acid (BHA) as the collector for rutile and sodium silicofluoride (SSF) as the inhibitor for garnet. Even with BHA having a good collecting performance for both rutile and garnet, there are still some differences. BHA greatly improves the hydrophobicity of rutile and garnet, and changes the chemical environment of rutile but not garnet. SSF significantly reduces the hydrophobicity of rutile and garnet, and slightly affects the environment in which BHA interacts with rutile. However, the above reagents and combinations have little effect on the surface chemical environment of garnet.

Keywords: flotation; rutile; fine particle; separation; collector selectivity

Citation: Zhang, Z.; Ren, L.; Zhang, Y.; Bao, S. Microflotation of Fine Rutile and Garnet with Different Particle Size Fractions. *Minerals* **2022**, *12*, 1238. https://doi.org/10.3390/min12101238

Academic Editors: Luis A. Cisternas and Dave Deglon

Received: 2 August 2022
Accepted: 26 September 2022
Published: 28 September 2022

1. Introduction

Garnet is the main associated gangue mineral of rutile, affecting its efficient recovery. The specific gravity of the two minerals is similar [1]. The iron isomorphism in the rutile lattice makes the surface of rutile easily contaminated with iron, and reduces the magnetic difference between rutile and garnet [2]. Therefore, flotation has become the main research direction for rutile and garnet separation.

Recently, there was renewed interest in the research on flotation separation of rutile and garnet, especially with respect to flotation reagents. Huang et al. [3] studied the flotation separation of rutile and garnet using the cationic surfactant octadecylamine polyoxyethylene ether as a collector, proving that the adsorption of AC1815 on rutile surface is mainly due to the electrostatic interaction and the hydrogen bonding; Richard et al. [2,4] studied the flotation separation of rutile and garnet using the cationic surfactant octadecylamine polyoxyethylene ether as the collector, Na–CMC as the inhibitor, sodium sulfite as a regulator, and octadecylamine polyoxyethylene ether-styrene phosphonic acid mixed as a composite collector, discovering that sodium sulfite acting in the form of SO_3^{2-} is more selectively adsorbed on garnet surface compared to that of rutile, leading to a high selectivity for the flotation of rutile; Chen et al. [1] investigated the flotation of rutile from garnet using sodium fluorosilicate as an inhibitor and styrylphosphonic acid as a collector, finding that there is a strong interaction between sodium fluorosilicate and active Fe sites on garnet surface, reducing adsorption sites for the collector, thereby strongly depressing the flotation of garnet. The above studies are mainly aimed at the mechanism of conventional particle size rutile and garnet flotation reagents, but the studies on the reagents for the flotation separation of fine particles of the two minerals still needs to improve.

According to all of the above researches, rutile and garnet can be effectively separated using phosphate, oleate, hydroxamic acid, and benzylarsinic acid as collectors, and sodium fluorosilicate, sodium hexametaphosphate, and Pb^{2+} as regulators. While carboxylic acid collectors have a strong collection performance, but poor selectivity, phosphonic acid collectors have a better flotation effect but a high cost; arsonic acid collectors have good collection performance and selectivity, but are highly toxic [5–8]. It was indicated that the selection of proper reagent is very important for separation [9,10]. The current research on flotation separation experiments pays more attention to hydroxamic acid collectors [11] because of their better selectivity, and lesser toxicity, while due to its low cost, styrene phosphoric acid is a common collector in industrial applications.

At present, the research on the flotation separation of rutile and garnet mainly focuses on the reagents [12,13], while there are fewer studies on fine particles and various reagent combinations. In addition, in actual flotation, due to factors such as disseminated mineral particle size and hardness, the particle size of target and gangue minerals may be different, while the research on the flotation separation of two minerals with different particle sizes is rare. Therefore, it is essential to further research the relation between inhibitor, activator, and collector, as well as their effects on the separation of rutile and garnet with different particle sizes.

This paper focuses on the effects of styrene phosphoric acid and hydroxamic acid collectors, sodium fluorosilicate inhibitor, and activator lead ions on the flotation of rutile and garnet. Referring to the properties of the Zaoyang rutile deposit in Hubei, China, -10 μm rutile and -30 μm garnet were selected as the research objects, which provided a certain basis for the flotation separation of rutile and garnet of different particle sizes in actual minerals.

2. Experimental Section

2.1. Sample and Reagents

The rutile and garnet samples used in this paper were from Hubei Province, China. According to the chemical analysis results of rutile in Table 1, the grade of rutile is 96%. The sample is of high purity from the X-ray diffraction analysis of the garnet sample in Figure 1. Both rutile and garnet samples meet the pure mineral flotation standard. According to the dissemination size of the two minerals in the rutile deposit in Zaoyang City, Hubei Province, the particle size fraction of rutile was selected to be -10 μm, and for garnet was -30 μm. After crushing with XPC-60 × 100 jaw crusher (Produced by China Henan Jiufu Machinery Equipment Co., Ltd., Zhengzhou, China), wet screening, and grinding with MJ-2A stirring mill at 500 rpm for 40 min, the D90 of the final rutile sample is 7 μm (-10 μm), and the D90 of the garnet sample is 25.38 μm (-30 μm).

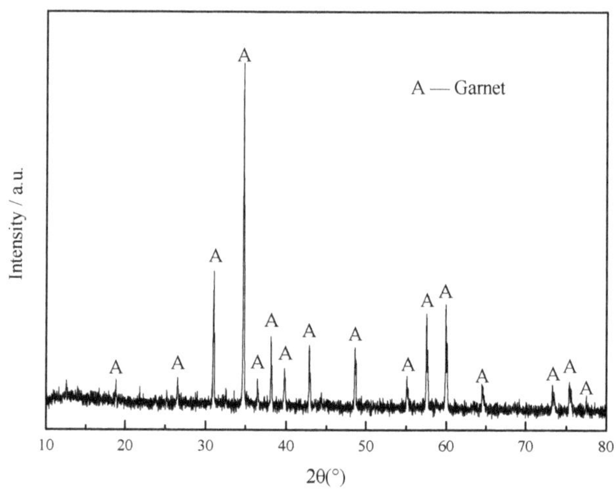

Figure 1. XRD pattern of garnet.

Table 1. Analysis results for the chemical composition of rutile/%.

TiO_2	ZrO_2	Nb_2O_5	SiO_2	Fe_2O_3	Al_2O_3	Cr_2O_3	CaO	ThO_2
96.21	0.73	0.61	0.51	0.44	0.31	0.22	0.19	0.19

The reagents used in this paper included benzohydroxamic acid (BHA), salicylic hydroxamic acid (SHA), styrene phosphoric acid (SPA), sodium fluorosilicate (SSF), $(NaPO_3)_6$, $Pb(NO_3)_2$, NaOH, H_2SO_4, and terpineol oil. All reagents except SPA were purchased from Shanghai Aladdin Reagent and all of analytical grade, while SPA was purchased from an industrial production site, and was technical grade pure. The test water was ultrapure, with a conductivity of 18.25 MΩ.cm.

2.2. Microflotation Tests

For microflotation tests, a XFGII flotation machine with a 50 mL cell was used at 1902 rpm. BHA, SHA, and SPA were used as collectors, SSF and $(NaPO_3)_6$ as inhibitors, Pb^{2+} as activator, and terpineol oil as frother. Except for terpineol oil, the reagents used at a concentration of 1 mg/mL 3 g of sample were examined in each experiment. Following pH adjustment (pH = 5~9), after each reagent addition (regulator, inhibitor, activator, collector), the mixture was stirred for three minutes. Afterwards, terpenic oil was added and the pulp was conditioned for 2 min, followed by 2 min flotation with the air flow rate of 0.1 m^3/h. The detailed flow of the flotation experiment is shown in Figure 2. The effects of collectors on rutile and garnet flotation were studied, then the effects of inhibitors under different collector conditions were explored, and finally, the effect of activators was determined.

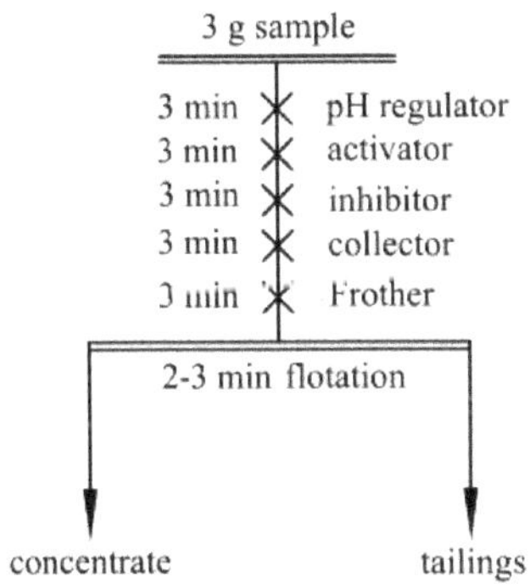

Figure 2. Flowsheet of flotation.

2.3. Zeta Potential Measurement

A nano-ZS90 zeta potentiometer was used to measure the electrophoretic mobility of rutile and garnet to examine the effect of different reagents on the electric surface potential of the two minerals. The test temperature was kept at 25 °C, the sample particle size was −5 μm, and the reagent concentration was 50 mg/L. After the reagent was added, the slurry was stirred for 15 min and then allowed to settle for 5 min, before the supernatant was taken for measurement. Each group of tests was performed in triplicate, and the average results were used for analysis.

2.4. FTIR Tests

A Nicolet 6700 Fourier transform infrared spectrometer was used to study the adsorption of the reagents on the surface of the two minerals, and the rutile and garnet samples before and after the conditioning with collectors, inhibitors, activators, and their combinations. A total of 1 g of the sample was used to prepare the slurry with the same concentration as the flotation concentration. After adding the reagent, it was stirred for

15 min, and then filtered and dried for testing. The adsorption performance of each reagent on the sample surface was compared and analyzed by the test results.

2.5. Contact Angle Measurements

To study the effect of the reagents on the contact angle of the two minerals, the rutile and garnet samples were conditioned with each reagent and then dried at low temperature, and each group of samples was pressed into a tablet shape by a tablet machine for testing. The contact angles of the rutile and garnet before and after each reagent's action were measured by a JC2000D. A droplet of 0.3 μL volume was produced and the stage carrying the tablet-pressed sample was raised to contact the droplet, and then lowered, and the contact angle of the droplet on the surface of the sample was measured. Each surface with or without reagent was tested 3 times.

3. Results and Discussion

3.1. Flotation Results of Different Types of Collectors

The effect of pH on flotation is shown in Figure 3. With BHA as the collector, the recoveries of rutile and garnet both reach their maximum at pH = 7, and simultaneously the largest difference (34.67%) between the recoveries of the two minerals is obtained. When using SHA as the collector, the recovery of garnet reaches its maximum at pH = 6, while that of rutile reaches its maximum at pH = 7. The difference between the recoveries of the two minerals at pH = 9 is larger, namely, 20.80%. With SPA as the collector, the recoveries of the two minerals reach their maximum values at pH = 8, and the difference of this is also the largest (21.95%). It is demonstrated that BHA probably works best for flotation separation of rutile and garnet in the above collectors.

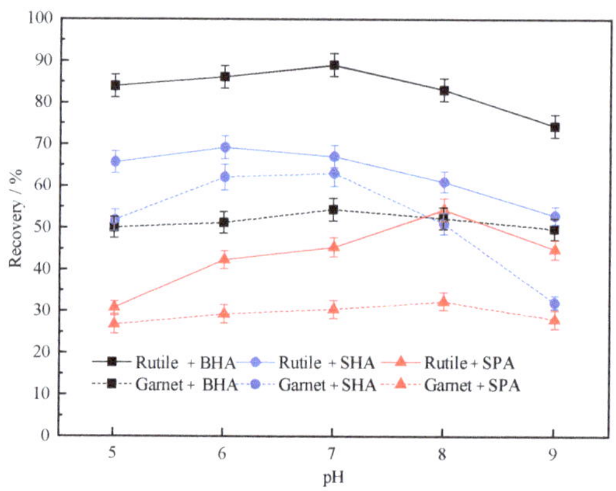

Figure 3. Effect of pH on the flotation of rutile and garnet with different types of collectors. (Concentration of collectors = 40 mg/L).

The effect of collector concentration on the flotation recovery is shown in Figure 4. With BHA as the collector, the increasing trend of the recoveries of rutile and garnet is linear with a low inclination when the concentration is 50 mg/L and 30 mg/L, and the recoveries are constant at 89.92% and 59.27%, respectively. When using SHA as the collector, the recoveries of rutile and garnet tend to be stable when the collector concentration is 40 mg/L, and are constant at 67.14% and 63.1%, respectively. With SPA as the collector, the recoveries of rutile and garnet also tend to be constant when the collector concentration is 40 mg/L, and are constant at 45.37% and 31.44%, respectively. In the BHA flotation system, the flotation recovery of the two minerals are maximum at different concentrations, which may be due to their selectivity to the two minerals resulting in more interaction sites with rutile [11].

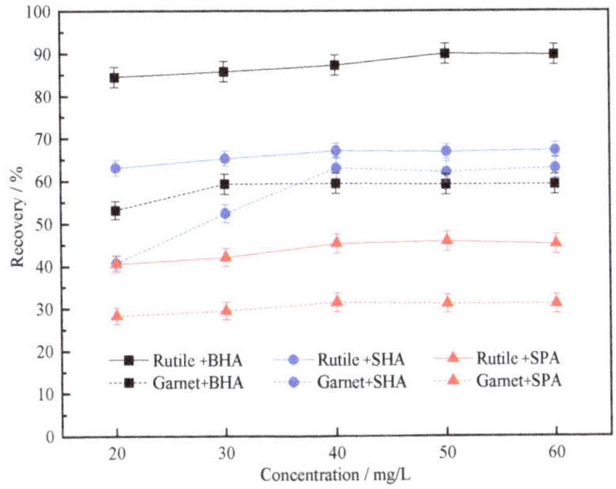

Figure 4. Effect of concentration on the flotation of rutile and garnet. (pH = 7 with BHA, pH = 9 with SHA, pH = 8 with SPA).

The selectivity and hydrophobization of BHA are better, while the hydrophobization and selectivity of SHA for fine particles are both poor, and the collecting ability of SPA was the worst, and the selectivity better than SHA, lower than BHA. With BHA as collector, the recovery difference between the two minerals is the largest when the concentration is 20 at pH = 7. However, it is difficult to achieve a great separation effect by using only a collector. Therefore, the effect of inhibitors on the flotation of two minerals was further studied.

3.2. Flotation Results of Inhibitors

Flotation results of inhibitors are as follows. Figure 5 shows the effect of pH on flotation under different inhibitor conditions. When SSF and BHA are used as reagents, the recoveries of rutile and garnet reach their maximum at pH = 7, and the difference between the recoveries of the two minerals is the largest at pH = 7, namely, 38.9%. The recovery of rutile in $(NaPO_3)_6$ and BHA flotation systems reaches the maximum at pH = 7, while the recovery of garnet reaches the maximum at pH = 8. The difference is the largest at pH = 7, but only 7.35%. $(NaPO_3)_6$ has a great inhibitory effect on the flotation of rutile, and has no obvious selectivity. This may be because $(NaPO_3)_6$ acts as a dispersant to disperse the fine rutile particles in the pulp, and it is difficult to achieve hydrophobic flocculation even under the action of the collector, which reduces the collision probability between the mineral particles and the bubbles, thus, affecting the floating effect.

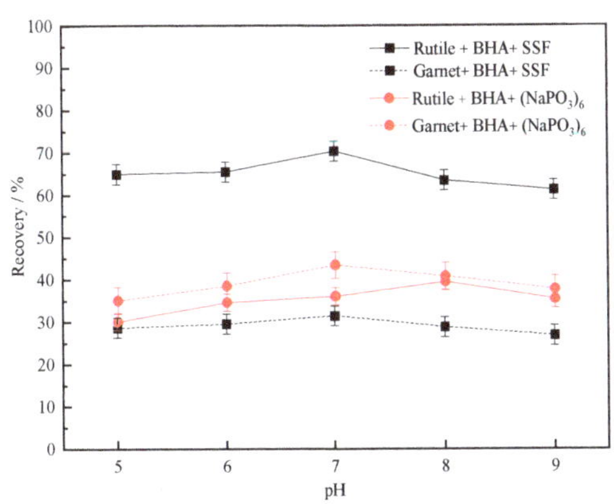

Figure 5. Effect of pH on the flotation of rutile and garnet by SSF and $(NaPO_3)_6$ (concentration of BHA = 20 mg/L, concentration of SSF = 10 mg/L, concentration of $(NaPO_3)_6$ = 10 mg/L).

Figure 6 shows the effect of inhibitor concentration on flotation. The recovery of the two minerals decreases with the increase in inhibitor concentration. In the SSF and BHA flotation systems, the recoveries of rutile and garnet are constant at 63.11% and 30.47% when the SSF concentration is 50 mg/L and 30 mg/L, respectively. In the $(NaPO_3)_6$ and BHA flotation systems, when the $(NaPO_3)_6$ concentration is increased to 30 mg/L and 50 mg/L, the recoveries of rutile and garnet are stable at 17.85% and 35.76%, respectively. The difference in inhibitor concentration for the two groups of minerals to achieve constant recovery may be due to the finer particle size of the rutile, the high specific surface area, and more action sites of the reagent. The inhibitory effect of $(NaPO_3)_6$ on rutile even exceeds that of garnet, proving that $(NaPO_3)_6$ is not available as an inhibitor for the flotation separation of rutile and garnet.

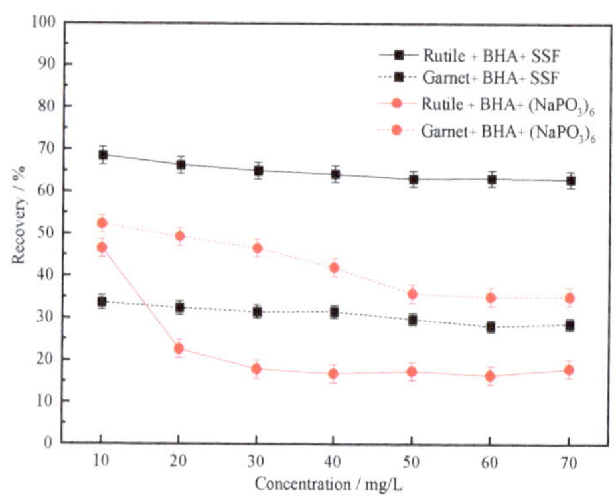

Figure 6. Effect of concentration on the flotation of rutile and garnet by SSF and $(NaPO_3)_6$ (pH = 7, concentration of BHA = 20 mg/L).

3.3. Flotation Results of Activator

According to the above experiments, BHA was selected as the collector and SSF as the inhibitor, and the effect of Pb^{2+} as the activator on the flotation behavior of rutile and garnet was studied. As shown in Figure 7, the flotation recovery for both minerals reaches maximum at pH = 7. The best recoveries of rutile and garnet are 72.51% and 49.44%, respectively, while without adding Pb^{2+}, the recoveries of the two minerals are 67.4% and 31.5%, respectively (Figure 6). Pb^{2+} can activate both rutile and garnet, but the difference between the two recoveries after activation is reduced to 23.07%, indicating that Pb^{2+} has a stronger activation effect on garnet and could not improve the separation effect of the two minerals.

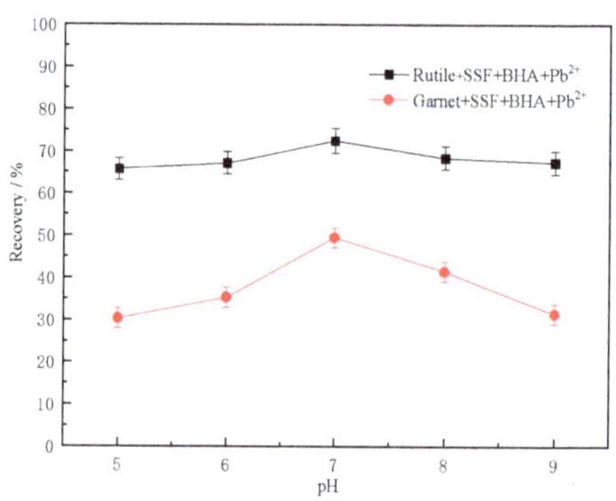

Figure 7. Effect of pH on rutile and garnet activated by Pb^{2+} (concentration of BHA and SSF = 40 mg/L, concentration of Pb^{2+} = 10 mg/L).

The effect of the concentration of Pb^{2+} on the flotation of rutile and garnet is shown in Figure 8. The recovery of both minerals increases with the increase in Pb^{2+} concentration. When the Pb^{2+} concentration is 15 μg/mL, the recovery values tend to be stable. The recovery of rutile increases from 65.96% to 72.74%, and the recovery of garnet increases from 31.4% to 45.95%, showing that the activation effect of Pb^{2+} on garnet is greater than that on rutile.

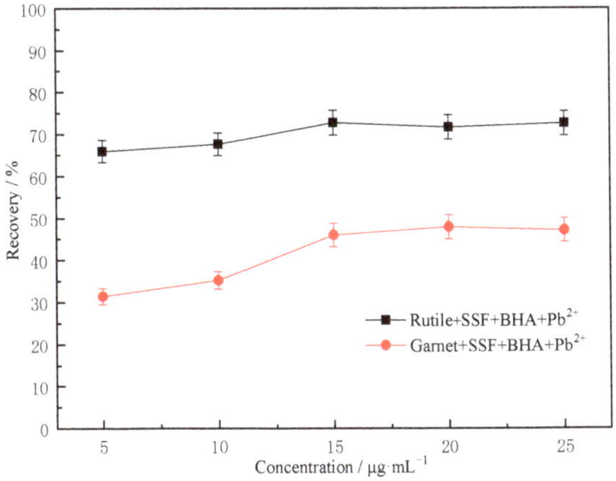

Figure 8. Effect of concentration on Pb^{2+} activation of rutile and garnet (pH = 7, concentration of BHA and SSF = 40 mg/L).

The addition of Pb^{2+} does not lead to better separation of garnet and rutile. Therefore, BHA was selected as the collector and SSF as the inhibitor, and the 1:1 artificial mixed ore of rutile and garnet was used for mixed flotation. The results of mixed flotation are shown in Figure 9. It can be seen that the index does not meet the separation standard, and the separation of the two is not achieved. This may be because the particle size of rutile and garnet is different, and the rutile particle size was too fine, meaning it was seriously entrained in the mixed flotation [14]. In addition, the use of terpineol oil, a foaming agent, may also increase the recovery of garnet in mixed flotation.

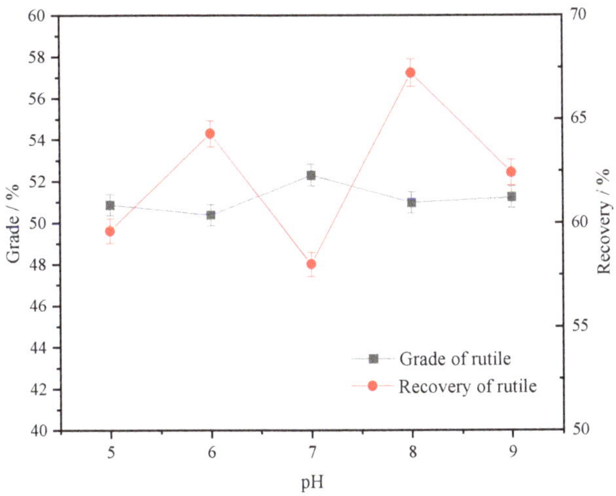

Figure 9. Mixed flotation results (concentration of BHA = 20 mg/L, SSF = 10 mg/L).

3.4. Contact Angle Measurements Results

Figure 10 shows the measurement results of the surface contact angle of rutile and garnet under the action of different reagents at pH = 7. The contact angles of pristine rutile and garnet samples are 32° and 24.5°, respectively. Compared to the contact angle of pristine mineral samples, the contact angle of rutile and garnet after the action of BHA is the largest, which is 50.5° and 38.5°, respectively, and the hydrophobicity is the maximum, followed by SHA, and SPA has the smallest change in the hydrophobicity of the two minerals. The contact angle difference between rutile and garnet after SHA is the largest, which is 12°, indicating that the selectivity of SHA is better. Among the two inhibitors, SSF has a greater effect on the contact angles of the two minerals, and significantly reduces their hydrophobicity, while the inhibitor $(NaPO_3)_6$ and the activator Pb^{2+} have few effects on the contact angles of the two minerals.

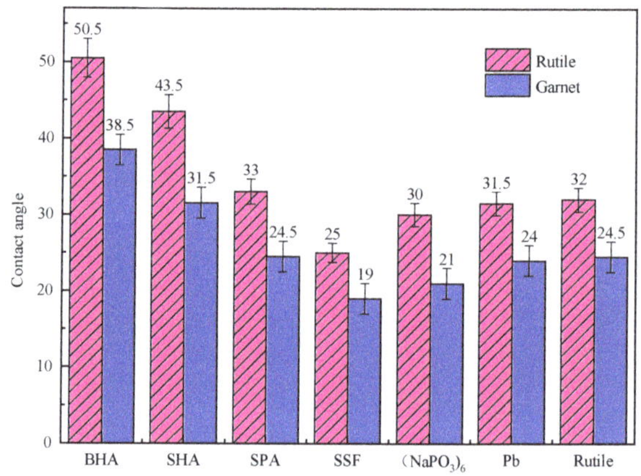

Figure 10. Effects of reagents on the hydrophobicity of rutile and garnet.

The effects of different chemical combinations on the surface hydrophobicity of rutile and garnet are shown in Figure 11. In the absence of the activator Pb^{2+}, the chemical combination of SSF and SPA has the greatest difference in the influence of the contact angle of rutile and garnet, which is more conducive to the separation of the two minerals. The contact angle of rutile and garnet after the action of BHA and SSF is the largest, and the hydrophobicity of minerals improves the most. With the addition of Pb^{2+}, the hydrophobicity of the two minerals does not change. This may be because Pb^{2+} affects the floatability of the minerals through other modes, which is also consistent with the results of the previous set of experiments and flotation experiments [8]. In addition, the results of the contact angle measurements may be smaller than that of the flaky mineral samples. This may be because the samples for the contact angle test are pressed from mineral powders, which affects the wettability of the samples under the action of capillary force, but has few effects on the study of the overall trend of hydrophobicity.

3.5. Reagent Adsorption Results

Figures 12 and 13 show the infrared test results of the interaction of reagents and minerals. After conditioning rutile with BHA, the adsorption bands appearing at 3434, 1714, and 1646 cm^{-1} are related to the stretching vibration of –OH, C=O, and the benzene ring in BHA [15,16]. The chemical environment of rutile is altered by BHA. Compared with the absorption bands 3434, 1714, and 1646 cm^{-1} under the action of BHA only, the absorption bands of each functional group under the action of BHA and SSF (3519, 1734, and 1643 cm^{-1}) have a slight shift and are very weak. This indicates that SSF affects the adsorption of BHA on rutile. Compared with the infrared spectrum of rutile treated with BHA, the absorption peaks of each functional group in the infrared spectrum of rutile under the action of Pb^{2+} and BHA show slight shifts. In addition, more complex absorption bands appear in the infrared spectrum, and the transmittance also changes, which indicates that Pb^{2+} has obvious changes on the chemical environment of rutile.

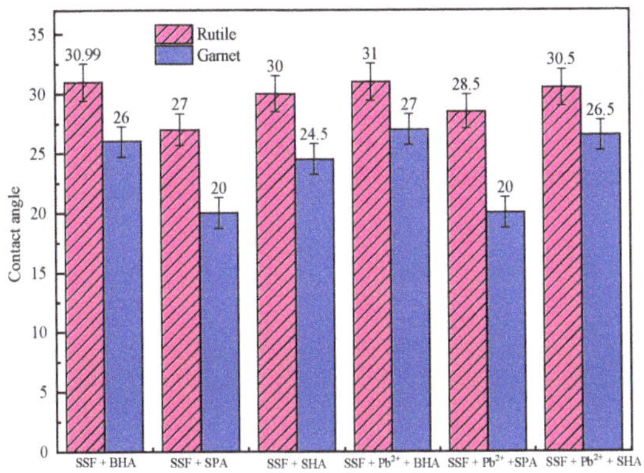

Figure 11. Effect of reagent combination on the hydrophobicity of rutile and garnet.

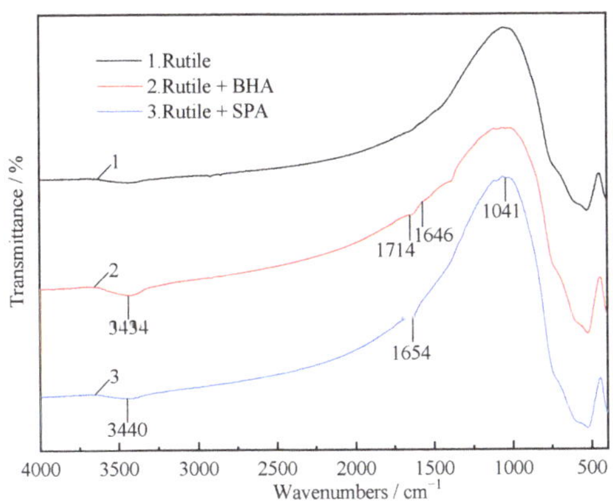

Figure 12. Infrared spectra of rutile before and after the interaction with reagents.

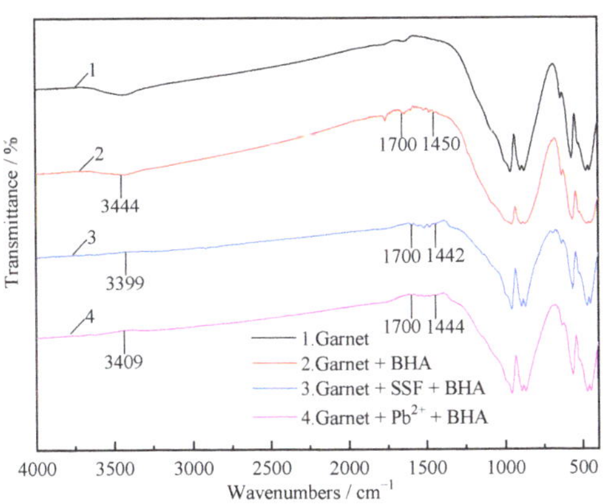

Figure 13. Infrared spectra of garnet before and after the interaction with reagents.

Figure 13 shows the adsorption results of reagent on the surface of the garnet. After conditioning with BHA, the absorption bands at 3444, 1700, and 1450 cm^{-1} have only slight changes in transmittance, which are related to the stretching vibration of OH, C=O, and benzene rings, respectively [17]. It suggested that there is little change in its chemical environment. Compared with the garnet treated with BHA, the infrared spectrum of the garnet treated with SSF and Pb^{2+} and then treated with collectors has only a difference in transmittance, and the change in transmittance is slight. It shows that both SSF and Pb^{2+} cause slight changes in the chemical environment of BHA, but due to the little environment alteration of BHA on garnet, the influence of inhibitors and activators is small.

3.6. Zeta Potential Measurement Results

The zeta potential test results of rutile and garnet before and after the action of each reagent are shown in Figure 14. The zeta potential of rutile shifts negatively with the increase in pH, because with the increase in OH^- concentration, the surface of rutile adsorbs OH^- to form a TiOH interface, which hydroxylates the surface of rutile [18]. When pH > 9, the adsorption of H^+ on the surface of garnet to the anionic active sites in the pulp exceeds the adsorption of the cationic active sites (Ti^+) and OH^-, causing the zeta potential to shift positively [19]. With the addition of BHA, the zeta potential of rutile shifts negatively, resulting in instability of the system, and the possibility of agglomeration of rutile particles. The garnet has a small negative shift, which may be because BHA does not change the chemical environment of garnet much and has little effect on zeta potential. With the addition of SSF, in comparison to the BHA action environment, the zeta potential of the rutile under the action of BHA shifts slightly positively, while that garnet shifts negatively. This indicates that either BHA and SSF compete for adsorption on the rutile surface, so the addition of SSF has little effect on the zeta potential of the rutile [20], or that SSF hinders the adsorption of BHA, leading to a lower negative shift in the zeta potential in the presence of both reagents. While SSF is adsorbed in garnet, the positively charged ions (the active site Fe^{2+}, Fe^{3+}, Al^{3+}) on its surface are adsorbed with negative ions in the solution, so that the surface of the garnet is negatively charged. SSF competes with anionic collectors for adsorption, which reduces the adsorption of BHA on the surface of the garnet, so SSF has an inhibitory effect on the garnet [21].

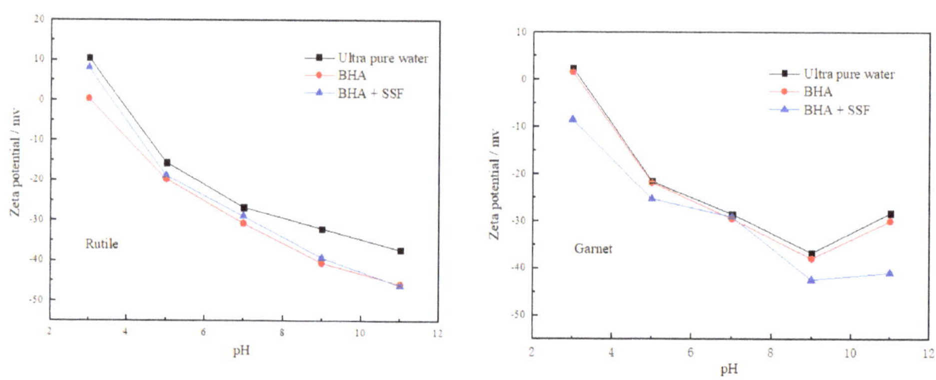

Figure 14. Effects of different reagents on the zeta potential of rutile and garnet.

4. Conclusions

(1) Among the collectors used in this paper, BHA has a better collection performance, and the difference in recovery between flotation rutile and garnet is the largest;

(2) SSF reduces the contact angles of rutile and garnet from 32° and 24.5° to 25° and 19°, respectively, significantly reducing their hydrophobicity. Among the collectors, BHA

increases the contact angles of the two minerals to 50.5° and 38.5°, which greatly improves their hydrophobicity;

(3) BHA has little effect on the surface potential of garnet, but negatively shifts the surface potential of rutile. The potential of rutile and garnet are adversely affected by the addition of SSF, and the effect on rutile is less;

(4) BHA and SPA change the chemical environment of rutile but not garnet. SSF slightly affects the environment in which BHA and SPA interact with rutile. However, various reagents and reagent combinations have little effect on the surface chemical environment of garnet. This provides a certain basis for the flotation separation of fine rutile and garnet, In the future, in addition to flotation reagents, the separation of the two can also be studied in terms of particle size.

Author Contributions: Conceptualization, Z.Z. and L.R.; methodology, Z.Z.; software, Z.Z.; validation, L.R., Y.Z. and S.B.; formal analysis, Z.Z.; investigation, Z.Z.; resources, Z.Z.; data curation, Z.Z. and L.R.; writing—original draft preparation, Z.Z.; writing—review and editing, Z.Z.; visualization, Z.Z.; supervision, L.R., Y.Z. and S.B.; project administration, L.R. and Y.Z.; funding acquisition, L.R. All authors have read and agreed to the published version of the manuscript.

Funding: This research was funded by National Natural Science Foundation of China grant number U2003129, 51504175, 52274270, and Natural Science Foundation of Hubei Province grant number 2019CFB533.

Data Availability Statement: Not applicable.

Acknowledgments: The authors gratefully acknowledge the National Natural Science Foundation of China (52274270, U2003129), and the Natural Science Foundation of Hubei Province (2019CFB533) for the financial support provided.

Conflicts of Interest: The authors declare no conflict of interest.

References

1. Chen, Q.; Tian, M.; Zheng, H.; Luo, H.; Li, H.; Song, S.; He, D.; Jiang, X. Flotation of rutile from almandine using sodium fluorosilicate as the depressan. *J. Sci. Colloids Surf.* **2020**, *599*, 124918. [CrossRef]
2. Richard, M.K.; Li, H.; Chen, Q.; Diallo, A.S.; Minani, I.; Weng, X.; Akisa, D.M.; Emmanuel, K.; Ge, W.; Song, S. Behavior and mechanism of sodium sulfite depression of almandinefrom rutile in flotation system. *J. Sci. Powder Technol.* **2020**, *374*, 49–57. [CrossRef]
3. Li, H.; Zheng, H.; Chen, Q.; Richard, M.K.; Leng, J.; Weng, X.; Song, S.; Xiao, L.; Tian, C. Flotation separation of rutile from almandine using octadecyl amine polyoxyethylene ether as collector. *J. Sci. Physicochem. Probl. Miner. Process.* **2020**, *56*, 653–664. [CrossRef]
4. Richard, M.K.; Li, H.; Chen, Q.; Weng, X.; Akisa, D.M.; Wu, S.G.; Song, S. Selective flotation of rutile from almandine using sodium carboxymethyl cellulose (Na-CMC) as a depressant. *J. Sci. Miner. Eng.* **2020**, *157*, 106544. [CrossRef]
5. Yu, X.; Cao, Q.; Zou, H.; Peng, Q. Activation mechanism of lead ions in the flotation of rutile using amyl xanthate as a collector. *Min. Metall. Explor.* **2019**, *37*, 333–344. [CrossRef]
6. Cao, M.; Gao, Y.; Bu, H.; Qiu, X. Study on the mechanism and application of rutile flotation with benzohydroxamic acid. *J. Sci. Miner. Eng.* **2019**, *134*, 275–280. [CrossRef]
7. Wang, J.; Cheng, H.; Zhao, H.; Qin, W.; Qiu, G. Flotation behavior and mechanism of rutile with nonyl hydroxamic acid. *J. Sci. Rare Met.* **2016**, *35*, 419–424. [CrossRef]
8. Xiao, W.; Zhao, H.; Qin, W.; Qiu, G.; Wang, J. Adsorption mechanism of Pb^{2+} mctivator for the flotation of rutile. *Minerals* **2018**, *8*, 266. [CrossRef]
9. Ren, L.; Qiu, H.; Zhang, Y.; Nguyen, A.V.; Zhang, M.; Wei, P.; Long, Q. Effects of alkyl ether amine and calcium ions on fine quartz flotation and its guidance for upgrading vanadium from stone coal. *Powder Technol.* **2018**, *338*, 180–189. [CrossRef]
10. Ding, W.; Bao, S.; Zhang, Y.; Xiao, J. Mechanism and kinetics study on ultrasound assisted leaching of gallium and zinc from corundum flue dust. *Miner. Eng.* **2022**, *183*, 107624. [CrossRef]
11. Zhang, Z.; Ren, L.; Zhang, Y. Role of nanobubbles in the flotation of fine rutile particles. *J. Sci. Miner. Eng.* **2021**, *172*, 107140. [CrossRef]
12. Wang, J.; Cheng, H.; Tan, W.; Hu, Y.; Jiang, T.; Liu, B.; Zhao, H.; Qiu, G.; Zhang, L.; Wang, Y.; et al. Rutile Beneficiation Method Comprising Magnetic Separation, Tailing Discarding Reselection: Desliming and Fine Particle Floatation. CN Patent 103586124 A, 19 February 2014.
13. Huang, X.; Xiao, W.; Zhao, H.; Cao, P.; Hu, Q.; Qin, W.; Zhang, Y.; Qiu, G.; Wang, J. Hydrophobic flocculation flotation of rutile fines in presence of styryl phosphonic acid. *J. Sci. Trans. Nonferrous Met. Soc. China* **2018**, *28*, 1424–1432. [CrossRef]

14. Li, H.; Afacan, A.; Liu, Q.; Xu, Z. Study interactions between fine particles and micron size bubbles generated by hydrodynamic cavitation. *J. Sci. Miner. Eng.* **2015**, *84*, 106–115. [CrossRef]
15. Köbl, J.; Fernández, C.C.; Augustin, L.; Kataev, E.Y.; Franchi, S.; Tsud, N.; Pistonesi, C.; Pronsato, M.E.; Jux, N.; Lytken, O.; et al. Benzohydroxamic acid on rutile TiO_2(110)-(1 × 1)– a comparison of ultrahigh-vacuum evaporation with deposition from solution. *J. Sci. Surf. Sci.* **2022**, *716*, 121955. [CrossRef]
16. He, J.; Sun, W.; Chen, D.; Gao, Z.; Zhang, C. Interface Interaction of Benzohydroxamic Acid with Lead Ions on Oxide Mineral Surfaces: A Coordination Mechanism Study. *J. Sci. Langmuir ACS J. Surf. Colloids* **2021**, *37*, 3490–3499. [CrossRef] [PubMed]
17. Zhang, C.; Zhou, Q.; An, B.; Yue, T.; Chen, S.; Liu, M.; He, J.; Zhu, J.; Chen, D.; Hu, B.; et al. Flotation Behavior and Synergistic Mechanism of Benzohydroxamic Acid and Sodium Butyl-Xanthate as Combined Collectors for Malachite Beneficiation. *Minerals* **2021**, *11*, 59. [CrossRef]
18. Bullard, J.W.; Cima, M.J. Orientation dependence of the isoelectric point of TiO_2 (rutile) surfaces. *J. Sci. Langmuir ACS J. Surf. Colloids* **2006**, *22*, 10264–10271. [CrossRef] [PubMed]
19. Yue, T. Flotation Separation of Fine-Grained Rutile. Ph.D Thesis, Northeastern University, Shenyang, China, 2006.
20. Dong, L.; Jiao, F.; Qin, W.; Zhu, H.; Jia, W. Selective depressive effect of sodium fluorosilicate on calcite during scheelite flotation. *J. Sci. Miner. Eng.* **2019**, *131*, 262–271. [CrossRef]
21. He, J.; Zhou, Q.; Chen, S.; Tian, M.; Zhang, C.; Sun, W. Interfacial microstructures and adsorption mechanisms of benzohydroxamic acid on Pb^{2+} activated cassiterite (110) surface. *J. Sci. Appl. Surf. Sci.* **2020**, *541*, 148506. [CrossRef]

minerals

MDPI

Article

Adsorption Difference of Octadecylamine on (002) and (131) Crystal Planes of Fine Muscovite and Its Guidance on Fine Muscovite Flotation

Liuyi Ren [1,2,*], Ziwei Jiao [1], Yimin Zhang [1,2,3,4] and Shenxu Bao [1,2]

1 School of Resources and Environmental Engineering, Wuhan University of Technology, Wuhan 430070, China
2 Hubei Key Laboratory of Mineral Resources Processing and Environment, Wuhan 430070, China
3 State Environmental Protection Key Laboratory of Mineral Metallurgical Resources Utilization and Pollution Control, Wuhan University of Science and Technology, Wuhan 430081, China
4 Hubei Collaborative Innovation Center for High Efficient Utilization of Vanadium Resources, Wuhan University of Science and Technology, Wuhan 430081, China
* Correspondence: rly1015@163.com or rly1015@whut.edu.cn

Abstract: Muscovite has a typical dioctahedral crystal structure. The atoms arranged in different directions of the crystal lattice will cause the anisotropy of the physical and chemical properties of the muscovite crystal planes, which also leads to the anisotropy of these crystal planes in flotation. In this study, Materials Studio 7.0 software was used to optimize muscovite crystal cells, and some properties of the (002) crystal plane and (131) crystal plane were calculated to obtain the anisotropy of these two crystal planes in some aspects, so as to further analyze their influence on flotation behavior. The differences of adsorption between these two crystal planes and octadecamine (ODA) were analyzed by XRD and flotation. The results showed that (002) has higher surface energy and was more easily exposed than (131). Compared with Si-O bond, Al-O bond in muscovite had lower covalent bond composition and was easier to break. O atom was the active site in the flotation of muscovite, and the (131) crystal plane was more likely to adsorb with ODA than (002) crystal plane. Through the simulation results, more (131) crystal planes can be exposed in the grinding stage to improve the flotation efficiency.

Keywords: muscovite; crystal plane; octadecylamine; flotation; adsorption; simulation

Citation: Ren, L.; Jiao, Z.; Zhang, Y.; Bao, S. Adsorption Difference of Octadecylamine on (002) and (131) Crystal Planes of Fine Muscovite and Its Guidance on Fine Muscovite Flotation. *Minerals* **2023**, *13*, 519. https://doi.org/10.3390/min13040519

Academic Editor: Luis Vinnett

Received: 20 March 2023
Revised: 3 April 2023
Accepted: 4 April 2023
Published: 6 April 2023

1. Introduction

Muscovite is a typical tetrahedron-octahedral-tetrahedral (TOT)-layered silicate mineral [1–3], usually in plate or sheet form. It has excellent dielectric properties and high thermal stability, and is widely used in construction, metallurgy, and cosmetics industries [4–8]. Flotation plays a very important role in muscovite beneficiation. Among many flotation agents, the interaction mechanism between cationic collector and muscovite is widely studied. It was concluded that the longer the carbon chain of alkyl primary amines is, the better the surface hydrophobicity of muscovite can be improved [9]. Many scholars have demonstrated this view by combining molecular dynamics simulations. Due to the differences in chemical and physical properties of muscovite crystals, such as surface electrical properties and wettability and adsorption properties [10–14], the anisotropy of each crystal plane of muscovite results in different consequences when it interacts with reagents.

Many complex studies have now become readily available using simulation methods. Materials Studio software enables the inexpensive simulations of certain properties of crystals, and molecular dynamics (MD) simulations [15–21] are a valuable tool for studying the adsorption of surfactants on solid surfaces at the microscopic level. The molecular dynamics simulation is based on density functional theory. After calculating the electronic properties of mineral crystals by density functional theory, the related properties of mineral crystals and their surfaces can be studied more simply and accurately on this basis, as well as the interaction mechanism between different crystal planes and chemical substances. Gao et al. [22] simulated the total energy of scheelite crystals and the surface energy of different surfaces. The results show that the order of surface energy of scheelite crystal planes is (001) < (112) < (111) < (110) < (100). According to the summary of the crystal anisotropy of phyllosilicate minerals and its relationship with flotation, the order of the fracture bond density of phyllosilicate minerals surface is $N_{Al\text{-}O}$ (100) > $N_{Al\text{-}O}$ (010) > $N_{Al\text{-}O}$(110), $N_{Si\text{-}O}$ (100) > $N_{Si\text{-}O}$ (010) > $N_{Si\text{-}O}$(110), and the order of hydrophobicity is (110) plane > (010) plane > (100) plane [23]. MD simulations are also used to simulate and calculate the fracture bond density and surface energy of the different crystal planes of smithite [24], so as to obtain the relationship between them. The simulation results show that the surface energy increases with the increase in fracture bond density. Not only that, MD simulations are also used [25] to research the adsorption equilibrium structure of octadecamine(ODA) on chlorite (001) and plane (100); the results show that chlorite (001) has a strong adsorption on ODA, while chlorite (100) has a weak adsorption on ODA. Although many scholars have simulated the adsorption mechanism of muscovite and agents, most of them focus on a common crystal surface of muscovite crystal, and few people have conducted comparative studies on the effect of a certain agent on different crystal surfaces of muscovite. Before flotation, different media or parameters can be selected for the grinding of muscovite samples [26], so that more crystal planes with strong adsorption of reagents can be exposed, thus improving the flotation efficiency of muscovite. In this study, the surface energy and fracture bond density of the (002) crystal plane and (131) crystal plane of muscovite were simulated by Materials Studio 7.0, and the properties of muscovite crystal were calculated to obtain the difference between the physical and chemical properties of these two crystal planes. The adsorption process of ODA on two crystal planes of muscovite was simulated by molecular dynamics and the differences in ODA adsorption between the two crystal surfaces were verified by flotation tests.

2. Theoretical and Experimental Sections

In the theoretical part, a series of operational processes for calculating muscovite using Materials Studio software are shown in Figure 1. For details, see Sections 2.1–2.3.

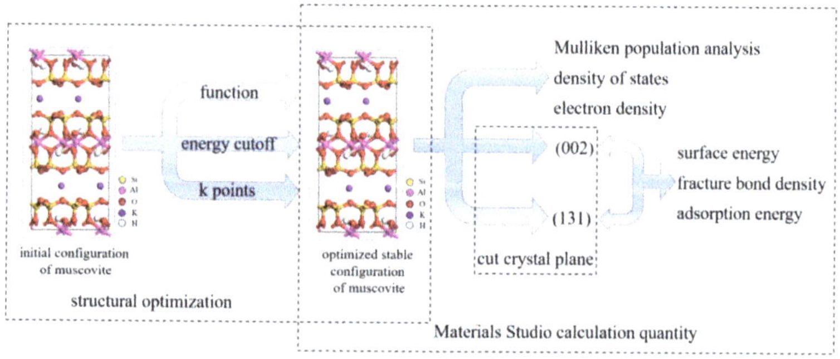

Figure 1. Flow chart of the theoretical part.

2.1. Models

The structural formula of muscovite is $KAl_2(AlSi_3O_{10})(OH)_2$, and its crystalline structure is a compound silico-oxygen layer with two silico-oxygen tetrahedrons sandwiched by an alumino-oxygen octahedron [27]. The crystal structure of monoclinic C2/C muscovite (Figure 2), which serves as the initial input structure for the MD simulation, was obtained by analyzing the single crystal diffraction data. Muscovite primitive cell parameters were as follows: a = 5.212 Å, b = 9.042 Å, c = 20.084 Å, alpha = 90°, beta = 95.78°, gamma = 90°.

ODA was used as the collector in the flotation test of muscovite. The molecular formula of ODA is $CH_3(CH_2)_{16}CH_2NH_2$. The optimized structure of ODA is shown in Figure 3.

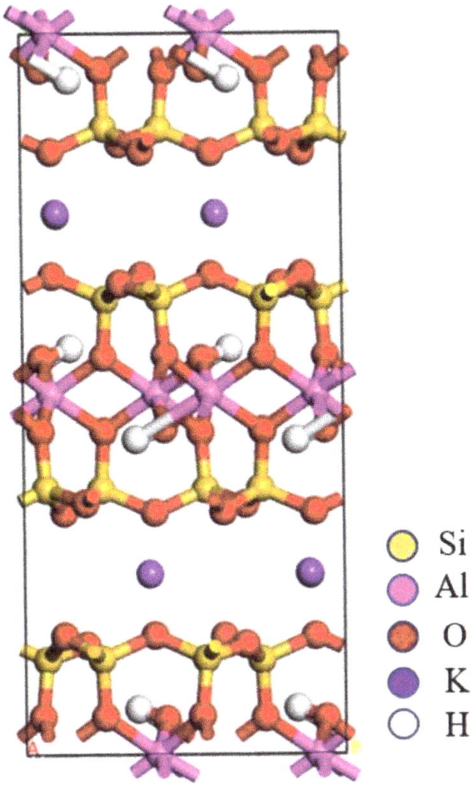

Figure 2. Original unit cell model of muscovite (the color of yellow, pink, red, purple, and white represents silicon, aluminum, oxygen, potassium, and hydrogen atoms, respectively).

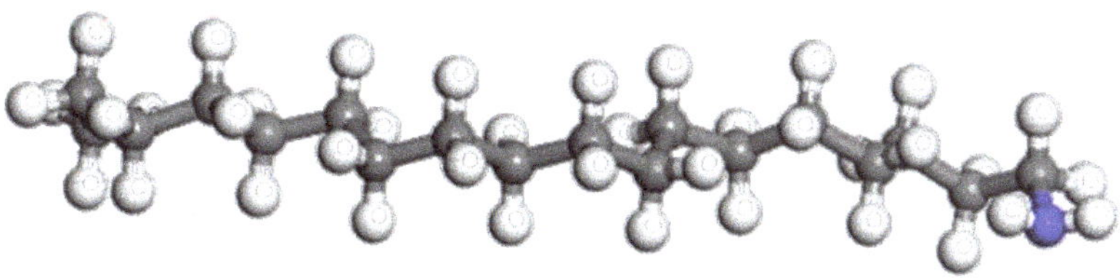

Figure 3. Molecular structure of octadecamine(ODA).

2.2. *Crystal Structure Optimization*

MD simulation was performed using Material Studio 7.0 software. Firstly, the CASTEP module was used to optimize the crystal structure of muscovite, and three parameters—function, energy cutoff, and k-points—were tested. The energy cutoff was 400 eV, the k-points were $3 \times 2 \times 1$, the mass was fine precision, and other parameters were not changed. For the generalized gradient approximation under the five density functions optimization calculation of muscovite lattice parameters, the results are shown in Table 1. It showed that when the function was GGA-PBESOL, muscovite phase had the lowest energy and the structure of muscovite was the most stable. Therefore, the function was selected as GGA-PBESOL.

Table 1. Lattice parameters and total energy of muscovite optimized by different functions.

Function	a/nm	b/nm	c/nm	β/°	Energy/eV	Maximum Error/%
GGA-PBE	5.241	9.073	20.934	95.423	−26,269.677	4.232
GGA-PBESOL	5.197	8.989	20.603	95.389	−26,315.327	2.584
GGA-WC	5.200	8.997	20.696	95.289	−26,228.042	3.047
GGA-PW91	5.232	9.058	20.764	95.430	−26,300.069	3.386
GGA-RPBE	5.287	9.162	21.292	95.667	−26,294.096	6.015

With other parameters unchanged, GGA-PBESOL was selected as the function, and different energy cutoffs were selected to optimize the calculation of muscovite cells. The cell parameters and total energy of muscovite obtained are shown in Figure 4. It shows that the muscovite phase always decreased with the increase in energy cutoff. When the energy cutoff was 550 eV, the total energy of muscovite reached the lowest value and then began to decline. When the energy cutoff was higher than 500 eV, the lattice constant error of muscovite also tended to be stable, at about 0.7, so the optimized energy cutoff of muscovite was selected as 550 eV.

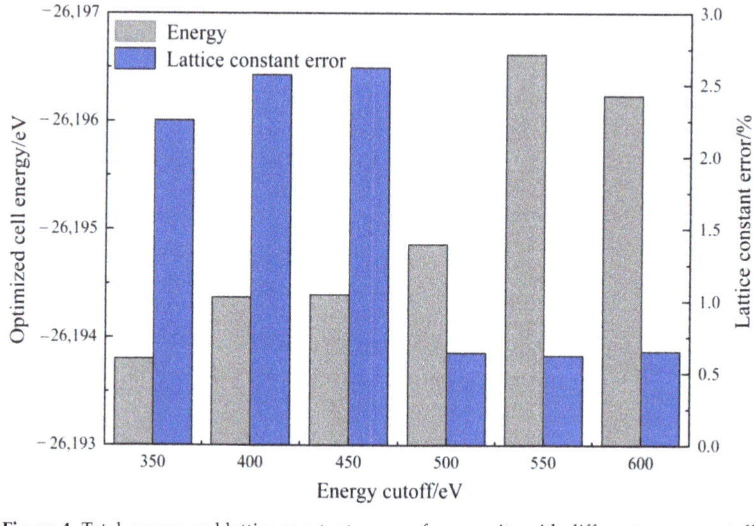

Figure 4. Total energy and lattice constant errors of muscovite with different energy cutoffs.

The function was selected as GGA-PBESOL, and the energy cutoff was 550 eV. The k-points convergence tests were conducted, and the obtained muscovite parameter error and total cell energy are shown in Figure 5. After determining the exchange correlation function and truncation energy, the change in k-points had little influence on the total energy of muscovite crystal. The energy changes were all within 0.1 eV, and the lattice constant errors were all less than 1%. When the k-points were $4 \times 3 \times 1$, the total energy of muscovite was the lowest, but when the k-points were $5 \times 3 \times 1$, the lattice error was only 0.642%, and the lattice constant was the closest to the experimental value. Therefore, the k-points chosen for calculating the relevant properties of muscovite were $5 \times 3 \times 1$.

Under these conditions, the calculated total energy of the system was −26,194.856 eV, and the error between the lattice constant and the experimental value was 0.642%. The calculation was reasonable. The method adopted and the parameters selected were reliable.

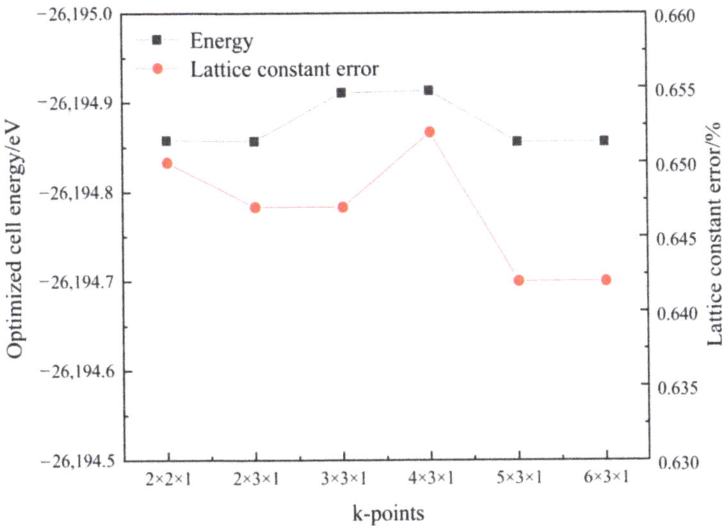

Figure 5. Total energy and lattice constant errors of muscovite at different k-points.

2.3. Calculation Quantity

2.3.1. Population Analysis

Mulliken population analysis is a data method for analyzing the extranuclear distribution of electrons in atoms, molecules, and crystals, which can be analyzed by atom population valence bond properties, orbital hybridization, and electron transfer in chemical changes [28]. The Mulliken population consists of two parts: the overlapping population and atomic population. The overlap population is a parameter to characterize bond covalence and the Mulliken charge in an atomic population is a parameter to characterize atomic ion properties.

The CASTEP module was used to calculate the performance of the optimized muscovite crystal structure. Properties were selected in the task bar for calculation, and population analysis was selected for specific properties. Other parameters remained unchanged.

2.3.2. Surface Energy and Fracture Bond Density

The number of unsaturated bonds per unit area is called the fracture bonds density [29], which can reflect the degree of mineral cleavage along the direction of a crystal plane. The lower the fracture bond density, the more easily the mineral cleavage along a crystal plane direction. Surface energy [30] is the energy required to break the crystal cleavage into two independent surfaces along the direction of a crystal plane under the action of external forces.

The surface energy and fracture bond density of muscovite were calculated using the CASTEP module. The surfaces were selected for cleaved surfaces in the build to cut the (002) and (131) crystal planes of muscovite. Before the calculation of surface energy, the convergence of layer thickness was tested and the appropriate thickness was selected for surface energy calculation. The calculation methods of fracture bond density and surface energy are shown in Formulas (1) and (2) [31,32].

$$D_b = N_b/S \tag{1}$$

$$E_{sur} = [E_{slab} - (N_{slab}/N_{bulk})E_{bulk}]/2A \tag{2}$$

where D_b is the number of unsaturated bonds in the area of 1 nm^2, N_b is the number of unsaturated bonds in the range of unit cell, and S is the area of unit cell. E_{sur} is the surface energy, E_{slab} represents the total energy of the surface structure, E_{bulk} is the total energy of the unit cell, N_{slab} is the number of atoms of the surface structure, N_{bulk} is the number of atoms of the unit cell, and A is the area of the surface structure along the Z axis.

2.3.3. State Density

State density represents the quantity of electrons allowed per unit energy range, that is, the distribution of electrons in a certain energy range. It shows how atoms interact with each other and how electrons are distributed in different orbitals. It also reveals information about chemical bonds. The CASTEP module of Materials Studio was used to select and calculate the state density of muscovite while conducting lattice optimization.

2.3.4. Electron Density

After the cell parameters and atomic coordinates of the crystal model of muscovite were optimized using the CASTEP module, the electron density of muscovite was analyzed. Its computational convergence accuracy was fine. Among electronic parameters, the sampling point setting (k-points set), SCF iteration error (SCF tolerance) and integration accuracy were all set to fine. The core treatment was set as all electron, the basis set as DND, and the atomic orbital radius (orbital cutoff quality) as medium.

2.3.5. Adsorption Energy

When one or more molecules move above the adsorption interface, a certain amount of energy will be released in the process of its velocity changing from large to small and finally will be adsorbed on the adsorption interface. The energy is called adsorption energy. The adsorption energy of different mineral surfaces interacting with the agent varies. As a result, the strength of the combination between the agent and different mineral surfaces can be judged by the adsorption energy. In the Forcite module, the adsorption configuration of ODA on the crystal planes of muscovite (002) and (131) were constructed and optimized. The adsorption energy calculation method of muscovite surface is shown in Equation (3) [33]:

$$\Delta E = E_{com} - E_{surf} - E_{rea} \tag{3}$$

where ΔE is the adsorption energy; E_{com}, E_{surf}, and E_{rea} are the optimized adsorbed complex, surface structure, and total energy of adsorbed substance, respectively. If the adsorption energy is spontaneous, the adsorption energy result is negative, and the smaller the adsorption energy is, the more stable the adsorption is. When the adsorption energy is 0 or positive, the adsorption cannot be spontaneous.

2.4. XRD

After weighing 0.5 g of muscovite raw ore and muscovite concentrate after flotation under different conditions, an XRD test was carried out to analyze the intensity changes in the (002) crystal plane and (131) crystal plane before and after flotation. The analysis was performed on a Netherlands Empyrean D8 Advance X-ray powder diffractometer with Kα radiation (λ = 0.15406 nm) at a scanning rate of 10°/min.

2.5. Flotation Test

An XFGC suspension flotation machine was used for a single mineral flotation. The mixing speed was kept at 2000 rpm. A total of 2 g muscovite was weighed for each experiment. After adding samples, the pulp was stirred for 3 min until fully mixed. The pH was adjusted and the pulp was stirred for 3 min; then, ODA was added and the pulp was stirred for another 3 min. Finally, the bubbles were aerated and scraped for 4 min. Finally, the concentrate and tailings obtained by flotation were dried, weighed, and the recovery was calculated.

The particle size of muscovite used for flotation is shown in Figure 6. The average of the sample is 4.02 um and P80 is 8.45 um. The particle density of the muscovite sample is 2.86 g/cm^3.

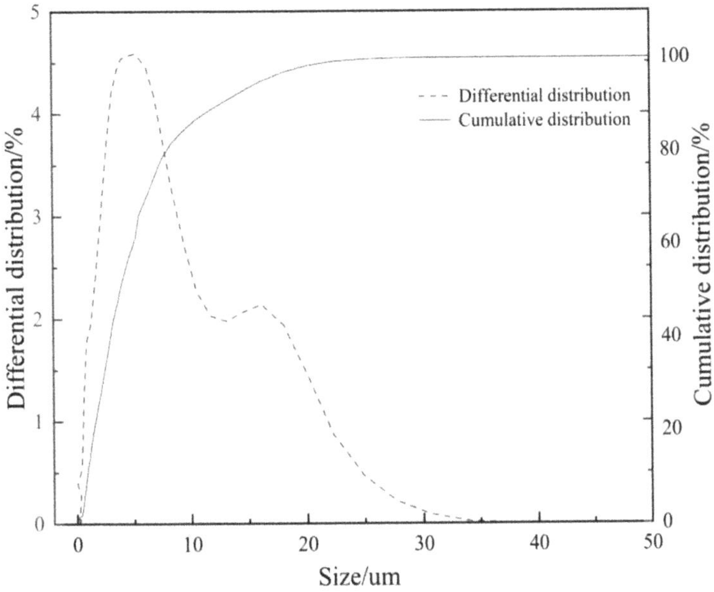

Figure 6. Grain size of muscovite.

3. Results and Discussion

3.1. Population Analysis and Electron Density of Muscovite

The population analysis of the optimized muscovite atoms is shown in Table 2, in which H, Al, and K atoms had two configurations, O atom had seven configurations, and Si atom had three configurations. Although there were many kinds of configurations, the charge changed little and the law was the same. H, Al, Si, and K atoms lost about 0.4, 1.7, 1.9, and 1.5 electrons, respectively, and were electron donors. O obtained about 1.1 electrons and was an electron acceptor.

Table 2. The Mulliken population analysis of muscovite atoms.

Species	Ion	s/e	p/e	d/e	f/e	Total/e	Charge/e
H	1	0.59	0.00	0.00	0.00	0.59	0.41
H	2	0.60	0.00	0.00	0.00	0.60	0.40
O	1	1.84	5.25	0.00	0.00	7.09	−1.09
O	2	1.85	5.29	0.00	0.00	7.15	−1.15
O	3	1.84	5.16	0.00	0.00	7.00	−1.00
O	4	1.86	5.30	0.00	0.00	7.16	−1.16
O	5	1.83	5.31	0.00	0.00	7.14	−1.14
O	6	1.84	5.29	0.00	0.00	7.13	−1.13
O	30	1.87	5.31	0.00	0.00	7.18	−1.18
Al	1	0.50	0.81	0.00	0.00	1.31	1.69
Al	2	0.47	0.85	0.00	0.00	1.33	1.67
Si	1	0.67	1.34	0.00	0.00	2.01	1.99
Si	5	0.68	1.36	0.00	0.00	2.04	1.96
Si	6	0.66	1.31	0.00	0.00	1.97	2.03
K	1	2.02	5.48	0.00	0.00	7.50	1.50
K	3	2.02	5.49	0.00	0.00	7.51	1.49

The ionic and covalent properties of the interatomic bonds of muscovite can be reflected by the Mulliken bond population value. If the population value is greater than 0, the atoms form bonds. The larger the population value, the more covalent components of the bonds, and the smaller the ionic bonds, the more components. Table 3 illustrates that only Al and Si atoms in muscovite bond with oxygen, and the population value between O and Si was 0.61, indicating more covalent components, while the population value between

O and Al was 0.30. More O-Al ionic bonding components compared to O-Si, so when muscovite was destroyed, O-Al was more likely to fracture than O-Si.

Table 3. The Mulliken bond population analysis of muscovite atoms.

Bond	Population	Length/Å
Al-Al	−0.37	2.9872
Si-Si	−0.17	2.9872
O-O	−0.03	2.9902
O-K	−0.04	2.9660
H-Si	−0.01	2.9480
Al-Si	−0.19	2.9379
H-Al	−0.02	2.8398
H-O	−0.01	2.8384
O-Al	0.30	1.9618
O-Si	0.61	1.5213

Figure 7 shows the electron density diagram of muscovite simulated by CASTEP. The blue part indicates that the atom lost electrons, while the grey part indicates that the atom gained electrons. As can be seen from Figure 7, in the alumino-oxygen tetrahedron and silico-oxygen octahedron of muscovite, only the oxygen atom has charge and presents a state of negative charge, indicating that the oxygen atom was a relatively active atom in muscovite.

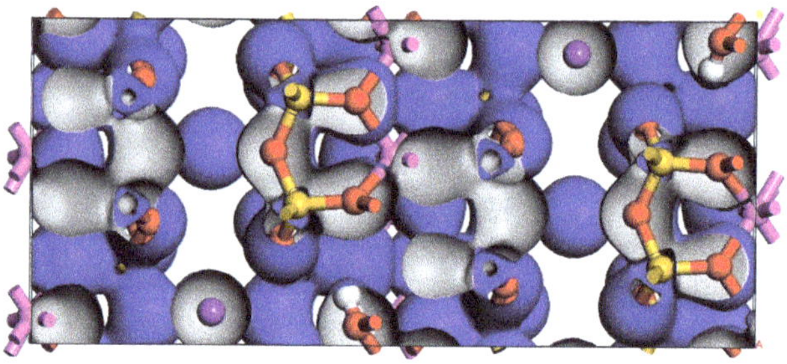

Figure 7. Electron density of muscovite (the color of yellow, pink, red, purple, and white represents silicon, aluminum, oxygen, potassium, and hydrogen atoms, respectively).

3.2. Anisotropy of Fracture Bond Density and Surface Energy at (002) and (131) Planes of Muscovite

After the convergence test of layer thickness, the atomic layer thickness of 3.877 nm and vacuum layer thickness of 1.400 nm were selected to calculate the surface energy of (002) crystal plane. For (131) crystal plane, an atomic layer thickness of 1.793 nm and vacuum layer thickness of 1.200 nm were selected to calculate the surface energy, as shown in Figure 8. The crystal plane models of (002) and (131) are shown in Figure 9, and the surface energy and fracture bond density calculated according to Formulas (1) and (2) are shown in Table 4.

The energy required to dissociate a mineral crystal along a crystal plane and split into two separate planes under the action of external forces is called surface energy. The smaller the surface energy, the more stable the crystal plane. Cleavage usually occurs along surfaces with relatively low surface energy. According to Table 4, the surface energy of the (002) crystal plane of muscovite was 1.349 J/m^2, lower than that of the (131) crystal plane, indicating that the (002) crystal plane was the normal texture plane of muscovite. There was a positive correlation between the fracture bond density and the surface energy of the crystal. The higher the fracture bond density was, the higher the surface energy was. According to the analysis of surface energy and fracture bond density, compared with the (131) crystal plane, the (002) crystal plane of muscovite was more easily exposed when damaged by an external force.

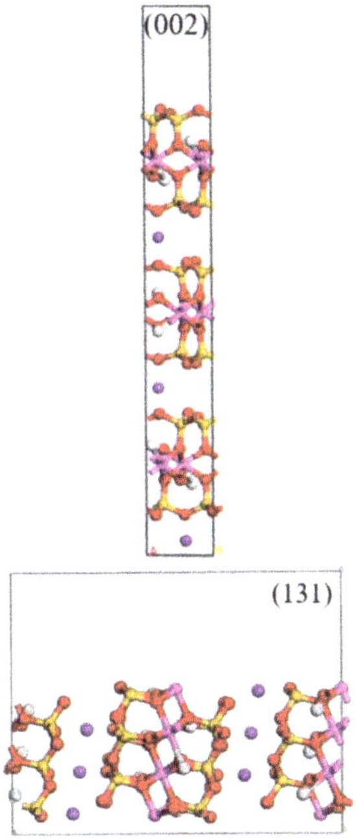

Figure 8. Surface energy calculation models of muscovite (002) and (131) crystal planes.

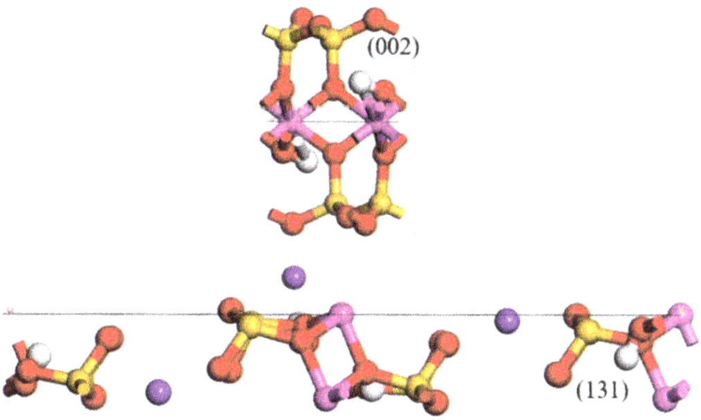

Figure 9. Calculation model of the fracture bond density of muscovite (002) and (131) crystal planes.

Table 4. Surface energy and the fracture bond density of muscovite (002) and (131) crystal planes.

Crystal Planes	Surface Energy/J·m^{-2}	Fracture Bond Density/nm^2
(002)	1.349	8.581
(131)	1.426	8.642

3.3. State Density of Muscovite

Figure 10 shows the total state density of muscovite crystal and the splitwave state density of each atom. It can be seen that each atom contributed to the total state density of muscovite. In combination with Figure 10b–d, and comparison with Figure 10a, the valence band is mainly composed of the p orbitals of oxygen atoms, and the orbitals of aluminum and silicon atoms contribute less. The conduction band was mainly composed of the p orbitals of silicon and aluminum atoms, and the s and p orbitals of oxygen atoms had little contribution. The main contribution of the adjacent region of the fermi level consists of the p orbital of the oxygen atom, so the oxygen atom had strong activity and was the active part of flotation beneficiation.

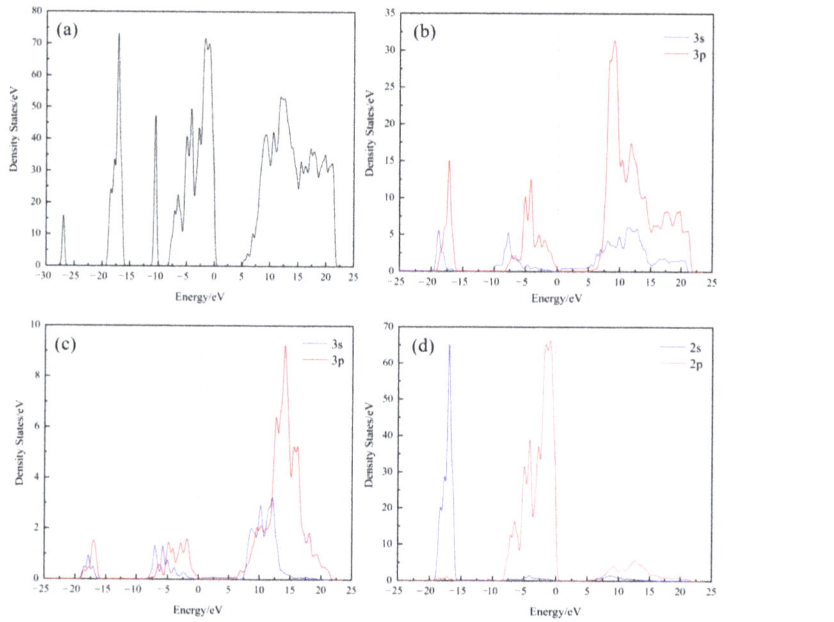

Figure 10. Muscovite state density ((**a**) total state density diagram (**b**) Al state density diagram (**c**) Si state density diagram (**d**) O state density diagram).

3.4. Adsorption Energy of Muscovite (002) and (131) Crystal Planes

The simulated adsorption energy results of the (002) crystal plane and (131) crystal plane with ODA are shown in Table 5. It indicates that the adsorption energy between the two crystal planes and ODA was negative, explaining that ODA can be adsorbed on both crystal planes, and the adsorption energy of the (131) crystal plane was much lower than the (002) crystal plane, showing that compared with the (002) crystal plane, the (131) crystal plane was more likely to adsorb ODA.

Table 5. Adsorption energy of muscovite (002) and (131) crystal planes.

Crystal Plane	Adsorbed Complex Energy /kJ·mol^{-1}	Energy at the Bottom of the Crystal Plane /kJ·mol^{-1}	Energy of Agent /kJ·mol^{-1}	Adsorption Energy/kJ·mol^{-1}
(002)	86,164.948	85,905.960	391.248	−132.259
(131)	49,649.868	50,185.722	22.052	−557.906

3.5. Adsorption Difference of ODA on (002) and (131) Crystal Planes of Muscovite

In order to obtain the adsorption morphology of ODA on two crystal planes of muscovite, the dynamic simulation of the interaction between the two crystal planes and the agent was carried out, and the adsorption changes of the agent on the crystal planes were obtained, as shown in Figures 11 and 12. After the adsorption of ODA on muscovite, the amine group head was closer to the crystal plane. The reason for this may be that the potassium ions between muscovite layers would lose part of the cleavage process, resulting in the electronegativity of muscovite (002) and (131) planes. The ammonium ion head group had positive charge and there was electrostatic attraction between it and the negatively charged muscovite surface. The cationic polar head group of the collector would approach muscovite and finally be adsorbed stably on the surface of muscovite. The (131) crystal plane might be more exposed to potassium ions, more electronegative, and easier to adsorb with ODA.

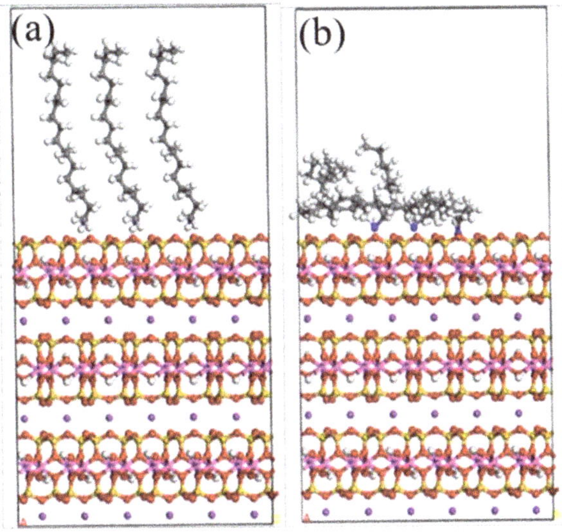

Figure 11. Changes in ODA before and after adsorption on the muscovite (002) crystal plane ((**a**) pre-adsorption and (**b**) post-adsorption).

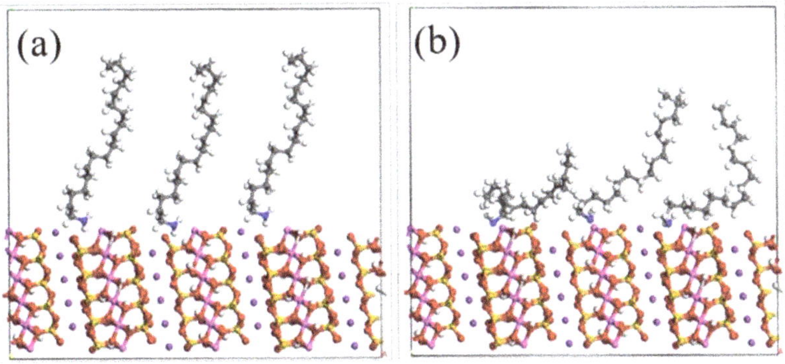

Figure 12. Changes in ODA before and after adsorption on the muscovite (131) crystal plane ((**a**) pre-adsorption (**b**) post-adsorption).

3.6. Flotation Test Results and XRD under Different Flotation Conditions

Figure 13 shows the XRD pattern of muscovite, and the crystal plane represented by its peak is labeled. Because the (002) crystal plane of muscovite was more exposed during grinding, the (002) crystal plane had a stronger peak. Among them, (002), (004) and (006) were parallel crystal planes with the same properties. Therefore, crystal planes (002) and (131) of muscovite were selected for subsequent simulation calculation to study the adsorption difference of ODA on these two planes.

When the concentration of ODA was 140 mg/L, under different pH conditions, the flotation recovery of muscovite changed. Figure 14 states that the recovery of muscovite increased significantly when the pH was from 2 to 4, and reached the peak value of 90.23% at pH 4. When the pH continued to rise, the flotation recovery began to decline [34].

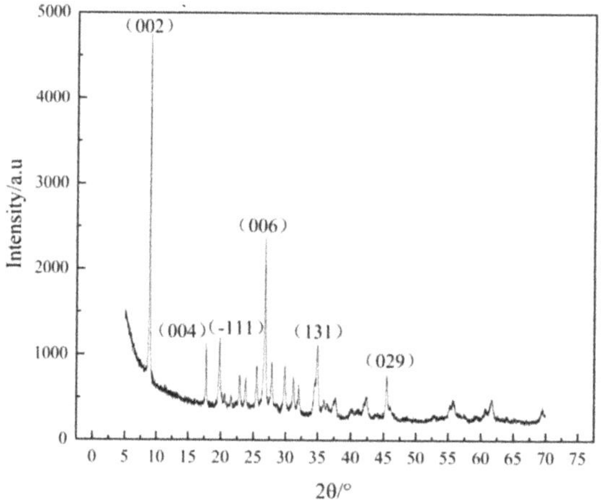

Figure 13. The XRD pattern of muscovite.

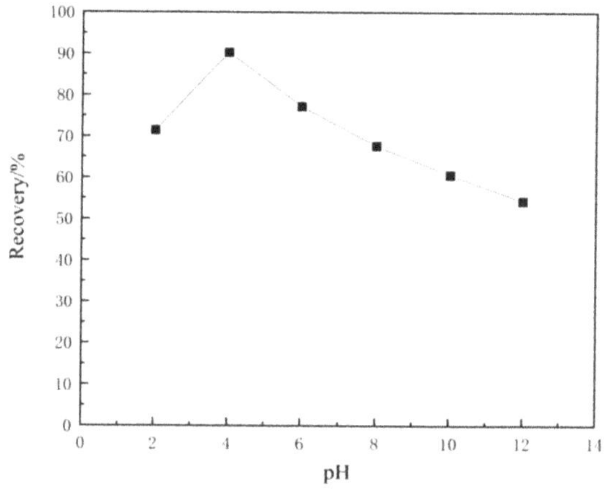

Figure 14. Relationship between the flotation recovery of muscovite and pH (concentration of ODA: 140 mg/L).

Figure 15 illustrates the effect of the ODA concentration on the recovery and contact angle of muscovite at pH 4. The recovery of muscovite increased with the increase in the ODA concentration [35]. In the range of low ODA concentration, the recovery of muscovite showed an upward trend, but when the ODA concentration reached 160 mg/L, the recovery of muscovite reached a relatively stable state. When the concentration of ODA was 180 mg/L, the recovery reached 96.55%. The contact angle of muscovite increased with the increasing concentration of ODA. When the concentration of ODA was 140 mg/L, the contact angle of muscovite was 50.5°. The increasing contact angle proved that the hydrophobicity of muscovite was increasing, which was consistent with the law of recovery.

Based on the previous research results, with a view to further investigate the influence of the difference of crystal plane properties on its flotation behavior, XRD tests were conducted on muscovite under different flotation conditions to observe the changes in the strength of crystal plane after flotation. The results are shown in Figure 16. It showed that the strength of the (002) crystal plane of muscovite decreased after flotation, but the degree of decline varied according to different flotation conditions. When the pH was 4, the decline of the muscovite (002) crystal plane slowed down with the increase in the ODA concentration. When the ODA concentration was 180 mg/L, the decline of the (002) crystal plane was the lowest. When the ODA concentration was 140 mg/L, pH increased from 4 to 8 and the decline of the (002) crystal plane decreased. When pH increased from 8 to 10, the decline of the (002) crystal plane increased. The strength of the muscovite (131) crystal plane increased after flotation, but the strength increased little with the change in conditions. After flotation by ODA, the change law of the two crystal planes of muscovite showed that the strength of the (002) crystal plane decreased and that of the (131) crystal plane increased as a whole, which proved that the adsorption of the (131) crystal plane was stronger than that of the (002) crystal plane during ODA flotation, which was consistent with the simulation results.

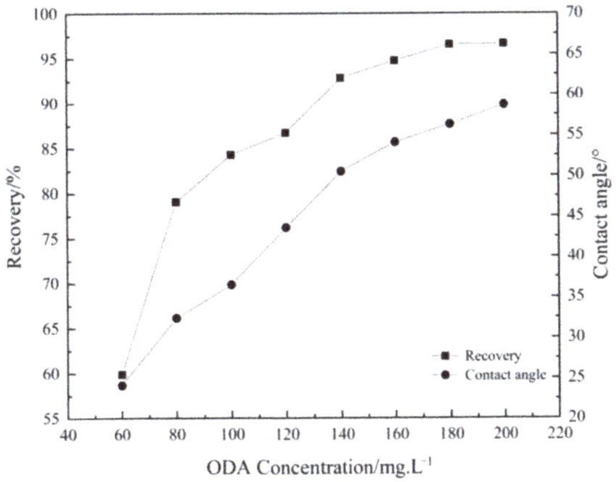

Figure 15. Flotation recovery and contact angle of muscovite in relation to ODA concentration (pH = 4).

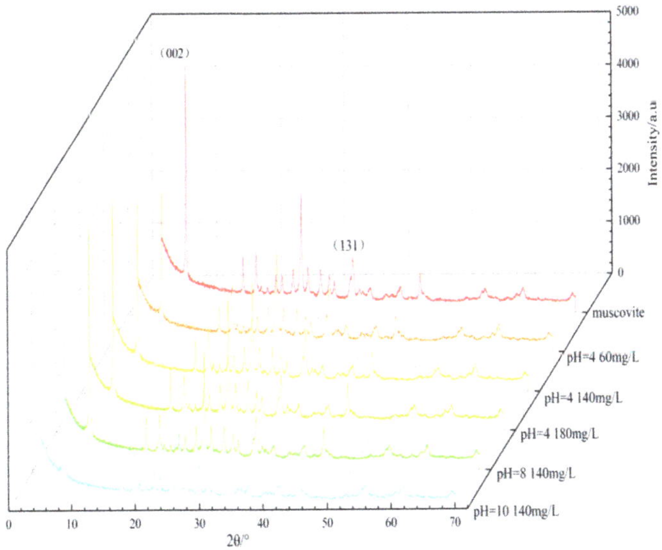

Figure 16. The XRD of muscovite (002) and (131) crystal planes under different flotation conditions.

4. Conclusions

Some properties of two crystal planes of muscovite (002) and (131) were simulated and explored, and the results showed that the (002) crystal plane had lower surface energy and fracture bond density. The adsorption energy of the (131) crystal plane was lower and more easily adsorbed ODA. Single mineral flotation tests were carried out on muscovite, and the optimum flotation conditions were determined as pH 4, ODA concentration 180 mg/L, and the recovery was 96.55%. The XRD results of the muscovite concentration under different flotation conditions showed that the strength of the (002) crystal plane decreased, while that of the (131) crystal plane increased. This was consistent with the simulation results, proving that in the flotation process, the (131) crystal plane was more likely to interact with ODA. This study has certain guiding significance for the flotation of muscovite, and can expose more of the (131) crystal plane before flotation to improve flotation efficiency.

Author Contributions: Conceptualization, L.R. and Z.J.; methodology, Z.J.; software, Z.J.; validation, L.R.; data curation, Z.J.; writing—original draft preparation, Z.J. and L.R.; writing—review and editing, Z.J., L.R. and S.B.; visualization, Y.Z. and S.B.; supervision, L.R.; project administration, L.R.; funding acquisition, L.R. All authors have read and agreed to the published version of the manuscript.

Funding: This research was funded by the National Natural Science Foundation of China (U2003129, 52274270).

Data Availability Statement: Data sharing not applicable.

Acknowledgments: The financial support provided by the National Natural Science Foundation of China (U2003129, 52274270) is acknowledged.

Conflicts of Interest: The authors declare no conflict of interest.

References

1. Zhou, S.; Lv, J.; Guo, L.K.; Xu, G.Q.; Wang, D.M.; Zheng, Z.X.; Wu, Y.C. Preparation and photocatalytic properties of N-doped nano-TiO_2/muscovite composites. *Appl. Surf. Sci.* **2012**, *258*, 6136–6141. [CrossRef]
2. Chen, S.; Chen, F.; Di, Y.; Han, S.; Zhu, X. Preparation and characterisation of exfoliated muscovite/poly(2,3-dimethylaniline) nanocomposites with an enhanced anticorrosive performance. *Micro Nano Lett.* **2020**, *15*, 509–513. [CrossRef]
3. Di, Y.; Jiang, A.; Huang, H.; Deng, L.; Zhang, D.; Deng, W.; Wang, R.; Luo, Q.; Chen, S. Molecular Dynamics Simulation Study on the Interactions of Mixed Cationic/Anionic Collectors on Muscovite (001) Surface in Aqueous Solution. *Materials* **2022**, *15*, 3816. [CrossRef] [PubMed]
4. Jiang, H.; Gao, Y.; Khoso, A.S.; Ji, W.; Hu, Y. A new approach for characterization of hydrophobization mechanisms of surfactants on muscovite surface. *Sep. Purif. Technol.* **2019**, *209*, 936–945. [CrossRef]
5. Wang, L.; Hu, Y.; Liu, J.; Sun, Y.; Sun, W. Flotation and adsorption of muscovite using mixed cationic-nonionic surfactants as collector. *Powder Technol.* **2015**, *276*, 26–33. [CrossRef]
6. Wang, L.; Sun, N.; Liu, J.; Tang, H.; Liu, R.; Han, H.; Sun, W.; Hu, Y. Effect of Chain Length Compatibility of Alcohols on Muscovite Flotation by Dodecyl Amine. *Minerals* **2018**, *8*, 168. [CrossRef]
7. Wang, L.; Sun, W.; Hu, Y.H.; Xu, L.H. Adsorption mechanism of mixed anionic/cationic collectors in Muscovite-Quartz flotation system. *Miner. Eng.* **2014**, *64*, 44–50. [CrossRef]
8. Chen, G.; Ren, L.; Zhang, Y.; Bao, S. Improvement of fine muscovite flotation through nanobubble pretreatment and its mechanism. *Miner. Eng.* **2022**, *189*, 107868. [CrossRef]
9. Liu, Z.; Liu, G.-S.; Yu, J.-G. Effect of Primary Alkylamine Adsorption on Muscovite Hydrophobicity. *Acta Phys.-Chim. Sin.* **2012**, *28*, 201–207. [CrossRef]
10. Gao, Z.; Li, C.; Sun, W.; Hu, Y. Anisotropic surface properties of calcite: A consideration of surface broken bonds. *Colloid Surf. A-Physicochem. Eng. Asp.* **2017**, *520*, 53–61. [CrossRef]
11. Wu, H.; Tian, J.; Xu, L.; Wang, Z.; Xu, Y.; Gao, Z.; Sun, W.; Hu, Y. Anisotropic surface chemistry properties of salt-type and oxide mineral crystals. *Miner. Eng.* **2020**, *154*, 106411. [CrossRef]
12. Xu, L.; Peng, T.; Tian, J.; Lu, Z.; Hu, Y.; Sun, W. Anisotropic surface physicochemical properties of spodumene and albite crystals: Implications for flotation separation. *Appl. Surf. Sci.* **2017**, *426*, 1005–1022. [CrossRef]
13. Xu, L.; Tian, J.; Wu, H.; Fang, S.; Lu, Z.; Ma, C.; Sun, W.; Hu, Y. Anisotropic surface chemistry properties and adsorption behavior of silicate mineral crystals. *Adv. Colloid Interface Sci.* **2018**, *256*, 340–351. [CrossRef]
14. Ren, L.; Zhang, Z.; Zeng, W.N. Adhesion between nanobubbles and fine cassiterite particles. *Int. J. Min. Sci. Technol.* **2022**, *in press*. [CrossRef]
15. Juarez, F.; Salmazo, D.; Savinova, E.R.; Quaino, P.; Belletti, G.; Santos, E.; Schmickler, W. The initial stage of OH adsorption on Ni(111). *J. Electroanal. Chem.* **2019**, *832*, 137–141. [CrossRef]
16. Loncaric, I.; Alducin, M.; Juaristi, J.I. Molecular dynamics simulation of O-2 adsorption on Ag(110) from first principles electronic structure calculations. *Phys. Chem. Chem. Phys.* **2016**, *18*, 27366–27376. [CrossRef] [PubMed]
17. Zhang, K.; Li, H.; Li, Z.; Qi, S.; Cui, S.; Chen, W.; Wang, S. Molecular dynamics and density functional theory simulations of cesium and strontium adsorption on illite/smectite. *J. Radioanal. Nucl. Chem.* **2022**, *331*, 2983–2992. [CrossRef]

18. Khnifira, M.; Boumya, W.; Attarki, J.; Mahsoune, A.; Abdennouri, M.; Sadiq, M.H.; Kaya, S.; Barka, N. Elucidating the adsorption mechanisms of anionic dyes on chitosan (110) surface in aqueous medium by quantum chemical and molecular dynamics. *Mater. Today Commun.* **2022**, *33*, 104488. [CrossRef]
19. Xu, J.; Yuan, Y.; Feng, Z.; Liu, F.; Zhang, Z. Molecular dynamics simulation of adsorption and diffusion of partially hydrolyzed polyacrylamide on kaolinite surface. *J. Mol. Liq.* **2022**, *367*, 120377. [CrossRef]
20. Yu, X.; Hu, X.; Zhao, Y.; Feng, Y.; Liu, J.; Dong, H.; Tang, H.; Wang, W.; Ren, W.; Wang, F.; et al. Molecular dynamics simulation of interface adhesion characteristics between dust suppressant and coal. *Mater. Today Commun.* **2022**, *33*, 104487. [CrossRef]
21. Hajjaoui, H.; Khnifira, M.; Soufi, A.; Abdennouri, M.; Akkaya, S.; Akkaya, R.; Barka, N. Experimental, DFT and MD simulation studies of Mordant Black 11 dye adsorption onto polyaniline in aqueous solution. *J. Mol. Liq.* **2022**, *364*, 120045. [CrossRef]
22. Gao, Z.Y.; Sun, W.; Hu, Y.H.; Liu, X.W. Surface energies and appearances of commonly exposed surfaces of scheelite crystal. *Trans. Nonferrous Met. Soc. China* **2013**, *23*, 2147–2152. [CrossRef]
23. Fang, S.; Xu, L.; Wu, H.; Tian, J.; Deng, W.; Chen, H. Research Advance in Anisotropy of Phyllosilicate Minerals and Its Relationship with Flotation. *Mod. Min.* **2018**, *34*, 46–53.
24. Wang, J.; Sun, Z.; Bai, J. Research on Crystal Anisotropy and Surface Properties of Smithsonite. *Conserv. Util. Miner. Resour.* **2021**, *41*, 1–6.
25. Li, M.; Lian, D.; Li, H.; Chen, W.; Hu, Y.; Gao, X.; Tong, X. Research on Anisotropy of Chlorite and Its Relationship with Floatability. *Met. Mine* **2020**, *2020*, 56–61.
26. King, R.P. *Butterworth-Heinemann Ltd. Modeling and Simulation of Mineral Processing Systems*; Society for Mining, Metallurgy & Exploration, Incorporated: Englewood, CO, USA, 2001; pp. 127–212.
27. Xue, X.; Xu, Z.; Pedruzzi, I.; Ping, L.; Yu, J. Interaction between low molecular weight carboxylic acids and muscovite: Molecular dynamic simulation and experiment study. *Colloid Surf. A-Physicochem. Eng. Asp.* **2018**, *559*, 8–17. [CrossRef]
28. Hasan, M.; Hossain, A. First-principles calculations to investigate the structural, electronic, optical anisotropy, and bonding properties of a newly synthesized ThRhGe equiatomic ternary intermetallic superconductor. *Results Phys.* **2022**, *42*, 106004. [CrossRef]
29. Kilic, K.I.; Dauskardt, R.H. Mechanically reliable hybrid organosilicate glasses for advanced interconnects. *J. Vac. Sci. Technol. B* **2020**, *38*, 060601. [CrossRef]
30. Meiser, J.; Urbassek, H.M. Influence of the Crystal Surface on the Austenitic and Martensitic Phase Transition in Pure Iron. *Crystals* **2018**, *8*, 469. [CrossRef]
31. Wu, Y.; Wang, Z.; Wang, D.; Qin, J.; Wan, Z.; Zhong, Y.; Hu, C.; Zhou, H. First-Principles Investigation of Atomic Hydrogen Adsorption and Diffusion on/into Mo-doped Nb (100) Surface. *Appl. Sci.* **2018**, *8*, 2466. [CrossRef]
32. Gao, Z.; Fan, R.; Ralston, J.; Sun, W.; Hu, Y. Surface broken bonds: An efficient way to assess the surface behaviour of fluorite. *Miner. Eng.* **2019**, *130*, 15–23. [CrossRef]
33. Gao, W.; Chen, Y.; Li, B.; Liu, S.P.; Liu, X.; Jiang, Q. Determining the adsorption energies of small molecules with the intrinsic properties of adsorbates and substrates. *Nat. Commun.* **2020**, *11*, 1196. [CrossRef] [PubMed]
34. Gomez, F.; Allan, H.; Graeme, W.; Sadia, K.; Hyunjung. Prediction of grade and recovery in flotation from physicochemical and operational aspects using machine learning models. *Miner. Eng.* **2022**, *183*, 107627. [CrossRef]
35. Kyuhyeong, P.; Junhyun, C.; Gomez-Flores, A.; Hyunjung, K. Flotation Behavior of Arsenopyrite and Pyrite, and Their Selective Separation. *Mater. Trans.* **2015**, *56*, 435–440. [CrossRef]

minerals

MDPI

Article

Frothing Performance of Frother-Collector Mixtures as Determined by Dynamic Foam Analyzer and Its Implications in Flotation

Khandjamts Batjargal [1], Onur Guven [2], Orhan Ozdemir [1,*], Stoyan I. Karakashev [3], Nikolay A. Grozev [3], Feridun Boylu [4] and Mehmet Sabri Çelik [4,5,*]

1 Department of Mining Engineering, Istanbul University-Cerrahpaşa, Buyukcekmece, 34452 Istanbul, Turkey
2 Department of Mining Engineering, Adana Alparslan Türkeş Science and Technology University, Sarıçam, 01250 Adana, Turkey
3 Department of Physical Chemistry, Sofia University, 1 James Bourchier Blvd, 1164 Sofia, Bulgaria
4 Department of Mineral Processing Engineering, Istanbul Technical University, Maslak, 34467 Istanbul, Turkey
5 The Institutes of Natural and Applied Sciences, Harran University, 63000 Şanlıurfa, Turkey
* Correspondence: orhanozdemir@iuc.edu.tr (O.O.); mcelik@itu.edu.tr (M.S.Ç.); Tel.: +90-4143183000 (M.S.Ç.)

Abstract: In recent years, most of the studies have been adapted to determine the optimum conditions for the flotation of very fine minerals. In this context, besides parameters such as particle size, morphology, and pH, the effects of frother type and its concentration present a very significant role in optimizing the flotation conditions. Therefore, the effects of froth stability during flotation can be considered one of the most important issues. Considering that knowledge in mind, in this study, the foamability and froth decay rate of six frothers (PPG200, PPG400, PPG600, BTEG, BTPG, and BDPG) having different molecular weights but similar polyglycol structures were investigated. In addition, methyl isobutyl carbinol (MIBC) which is a well-known frother type in the industry was also used as a reference. Additionally, a series of tests were also performed in the presence of collectors (Dodecylamine hydrochloride, DAH, and sodium oleate, NaOL) + frother mixtures to mimic the flotation conditions. The results of these tests indicated that the bubble size became finer at even low concentrations of PPG600 and PPG400 frothers. Following that, a significant decrease in bubble size was also found for the collector + frother mixtures system regardless of the concentration of the frothers.

Keywords: frothers; dynamic froth stability; collectors

Citation: Batjargal, K.; Guven, O.; Ozdemir, O.; Karakashev, S.I.; Grozev, N.A.; Boylu, F.; Çelik, M.S. Frothing Performance of Frother-Collector Mixtures as Determined by Dynamic Foam Analyzer and Its Implications in Flotation. *Minerals* **2023**, *13*, 242. https://doi.org/10.3390/min13020242

Academic Editors: Liuyi Ren, Wencheng Xia, Wei Xiao and Siyuan Yang

Received: 21 December 2022
Revised: 29 January 2023
Accepted: 30 January 2023
Published: 8 February 2023

1. Introduction

Flotation of fine-size minerals has become particularly important in recent years, as advances in milling have allowed the economical use of low-grade ore deposits [1]. Also, mineral wastes, mostly composed of fine particles, are becoming a critical problem for the mining industry due to environmental concerns (for instance, acid mine drainage) [2,3]. Accordingly, the flotation of fine particles depends on the recovery of valuable hydrophobic mineral particles from the pulp by adhering to fine-sized air bubbles. Thus, low recovery values of fine particles by flotation can be attributed to the low probability of bubble-particle collision, which decreases with decreasing particle size [4]. However, bubble-particle interactions such as electrostatic and hydrophobic forces are important in determining the selectivity of the flotation separation [5]. Frothers and surface–active collectors are often used for the flotation of different minerals [6,7]. Both of them often exhibit synergy in particle recovery but the possible synergetic effects on the froth stability have not been studied.

Derjaguin et al. (1984) [8] formally proposed that the flotation general rate constant is equal to the product of three possibilities or efficiencies: Bubble-particle collision, adhesion, and stability. The flotation rate constant, and hence the flotation recovery, is a function of the probability of these subprocesses [9]. Forty years after Derjaguin and Dukhin's proposal, Miettinen et al. (2010) [10] published a review on the flotation limits of fine particles, showing the importance of key parameters such as bubble size, particle aggregation, and flow conditions in fine particle flotation. Moreover, in a recent paper of ours [5] we proved theoretically that the collision rate between the fine particles and the bubbles is sufficient [11,12] to sustain high recovery if all of the collisions are effective. In addition, there exists a thermodynamic low size limit for the flotation of fine hydrophobized particles ($\theta = 80°$). This limit is in the range of 0.15–0.4 μm) and below it the capture of the fine particles is impossible. Particles with sizes in the range of 0.4–2 μm can be floated but with a small flotation rate due to the very small level of exceeding the total push force over the resistance force for the formation of a three-phase contact line (TPCL). Such particles with sizes above 2 μm are floated due to the cavitation of dissolved gases. So, there exist two ways to float fine particles with high recovery for short time—(i) to initiate electrostatic attraction between bubbles and particles; (ii) to hydrophobize them significantly ($\theta > 90°$). The second way is more powerful.

Nevertheless, improving the recovery of fine-sized particles (−20 μm) has been a long-standing goal in the mineral processing industry. The relationship between particle size and buoyancy is presented in detail by research conducted by Gaudin et al. (1931) [13], which shows that fine particles are more difficult to recover than medium-size particles. As the particle size decreases, the effectiveness of the bubble–particle collisions reduces. Introducing fine bubbles increases the fine particle recovery due to (i) increasing the number of bubbles in the flotation reactor, thus increasing the collision rate; (ii) larger capillary pressure during the bubble–particle collision due to the increased pressure inside of the fine bubbles. The latter factor is the powerful driving force for the capture of the particles by the bubbles.

Such methods for the production of small bubbles have been the focus of many studies including dissolved air flotation [14] and pico bubbles [15]. Because the problem of dealing with fine particles has preoccupied mineral flotation engineers for more than a century. This issue is even more important today, driven by the energy and water minimization requirements as well as the mineralogy and liberation requirements of many ore deposits. Considering that knowledge in mind, it is the objective of this study to systematically determine the froth stability characteristics of seven commercial frothers by the "Dynamic Froth Analysis" method as a function of frother and collector concentration.

2. Materials and Methods

2.1. Materials

In this study, the foamability and the rate of froth decay for the selected frothers obtained from BASF, Ludwigshafen, Germany, namely, polypropylene glycols (PPG200, 400, and 600), tri propylene glycol (BTPG), triethylene glycol (BTEG), dipropylene glycol (BDPG), and isobutyl carbinol (MIBC) was determined using a dynamic foam analyzer. In addition to frothers, the effects of collectors, namely, dodecylamine hydrochloride (DAH, >98% purity, Acros Organics) and sodium oleate (NaOL, >99% purity, Sigma-Aldrich) were also investigated as collector + frother mixtures in certain ratios (see Table 1). Some chemical and physical properties of frothers and collectors are presented in Tables 1 and 2, respectively.

Table 1. Some chemical and physical properties of frothers.

Chemical	Hydroxyl Value (mgKOH/g)	Molecular Weight (g/mol)	Formula	HLB	pH (1% Solution)	Molecular Stucture
PPG200 (n~3.5)	510~623	180~220	$HO(C_3H_6O)nH$	10.55	5.0~7.0	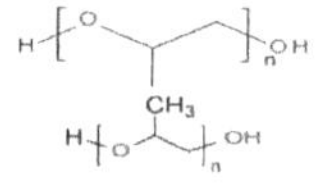
PPG400 (n~6.5)	255~312	360~440	$HO(C_3H_6O)nH$	8.69	5.0~7.0	
PPG600 (n~10)	170~208	540~660	$HO(C_3H_6O)nH$	8.25	5.0~7.0	
BTPG (n = 3)	~75 Polyglycols	248.36	$C_4H_9(C_3H_6O)nOH$	6.63	5.0~7.0	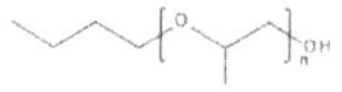
BTEG (n = 3)	≥50%–<100% Polyglycols	206.2793	$C_4H_9(C_2H_4O)nOH$	8.05	5.0~7.0	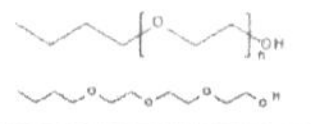
BDPG (n = 2)	≥98.5 Polyglycols	190.28	$C_4H_9(C_3H_6O)nOH$	6.75	5.0~7.0	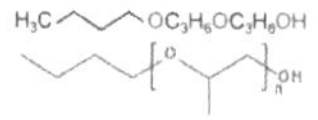
MIBC	*w*/*w*: ≥100%	102.17	$CH_3CH(OH)CH_2–CH(OH)CH_3$	6.53	5.0~6.0	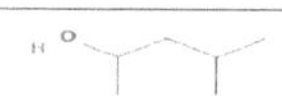

Table 2. Physical and chemical properties of collectors.

Chemical	Form	Purity (%)	Formula	Molecular Weight (g/mol)	CMC (mol/L)	Density (g/cm^3)	Molecular Structure
DAH	solid	≥99%	$C_{12}H_{28}ClNa$	221.81	$1.5{\cdot}10^{-2}$ (25 °C)	1.0 (20 °C)	
NaOL	solid	≥97%	$C_{18}H_{33}NaO_2$	304.44	$3{\cdot}10^{-3}$	0.9	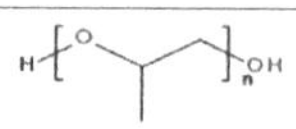

2.2. Methods

2.2.1. Dynamic Foam Stability Experiments

The foamability experiments were carried out using the KRUSS dynamic foam analyzer DFA100 (KRUSS GmbH, Hamburg Germany) shown in Figure 1. This device measures the kinetics of foam formation, and froth decay. An optical sensor is attached to the foam column and measures the discharge of liquid (foam drainage) from the foam lamellas and the amount of foam produced. The foam generator assembly is connected to a computer serving as a data collection and monitoring unit. All measurements, result evaluation, and analysis are controlled by the installed foam analysis software. Statistical accuracy was ensured by measuring the results at least three times.

Geldenhuys et al. (2018) [16] reported that pore opening influences dynamic foam stability, but the results should be comparable to each other as the pore opening is constant for all suspensions tested. The foam is then created by bubbling air through a ceramic frit at an airflow rate for a constant time (standard conditions). Its height is recorded to evaluate the dynamic parameters of foaming ability, the rate of foam decay, and time-dependent bubble size distribution.

For the dynamic foam stability experiments, 10,000 ppm (1 g/100 mL) stock solutions were first prepared for each frother. Then, standard solutions at CCC values for each frother were prepared. All these solutions were prepared by mixing distilled water (23° ± 1) at room temperature and 500 rpm mixing speed for 5 min. As a collector, 7 solutions were prepared in the concentration range of 1.10^{-5}–1.10^{-3} mol/L DAH. Similar to DAH, 7 concentrations of NaOL between 1.10^{-5} M–1.10^{-3} mol/L were prepared with distilled water (23° ± 1) as 100 mL for NaOL. In addition to CCC of frothers with the collectors 5.10^{-5} mol/L mix system. Concentrations of 1.10^{-5}–1.10^{-3} mol/L of collectors were used in the presence of frother PPG600 3 ppm. The experiments were carried out depending on the mixed solution concentrations of the collectors and frothers given in Tables A1–A7.

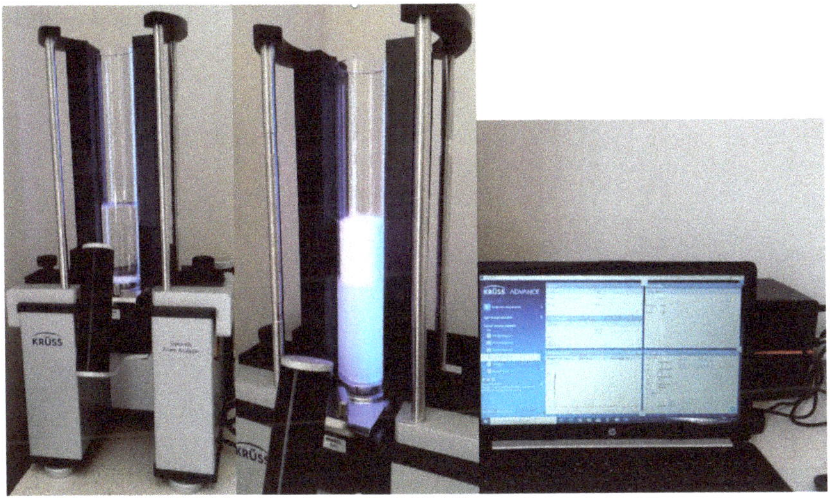

Figure 1. Set up for dynamic foam stability experiments using the Dynamic foam analyzer (DFA100).

In the experiments, 100 mL of solutions is first transferred into a tempered glass column with a diameter of 40 mm and a height of 250 mm. Then, when the device is started, the air is distributed through a porous filter plate (12–25 μm) in the system (from electronic gas flow control) at an airflow rate of 0.2 L/min (Figure 2a), resulting in the desired foam formation. The foam formation time is set at 60 s and the foam height in the system is measured during this time (Figure 2b). Then, the gas flow is stopped automatically at 60 s, and the foam degradation is measured for 40 s (Figure 2c).

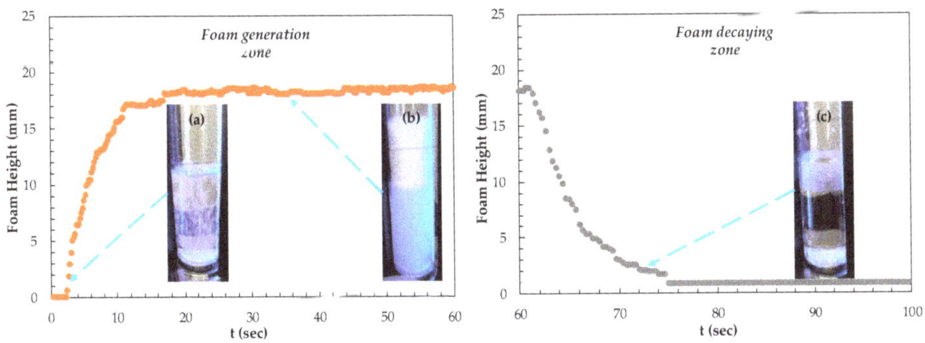

Figure 2. Dynamic foam stability experiments (**a**) foam formation (**b**) stable foam (**c**) foam decaying.

Our previous study indicated that there was not much difference between the air and N_2 gases in the experiments performed with a 10 ppm concentration of PPG600 frother depending on the time in terms of foamability [17]. For this reason, the air was used in the measurements.

2.2.2. Modeling Studies

The foaminess and foam decay curves were processed, thus calculating the Bickerman unit of foaminess [18] and foam production [19]. The Bickerman unit of foaminess is the ratio between the height of the stationary froth column and the superficial gas flow given through the porous bottom:

$$\sum = \frac{h_{foam}}{U_{gas}} \tag{1}$$

where h_{foam} is the stationary height of the foam, while is the gas flow rate (cm/s).

The foam production is the ratio between the Bickerman unit of foaminess and the initial speed of froth decay after stopping the gas flow:

$$FP = \frac{\Sigma}{\frac{dh}{dt}\Big|_{initial}} \tag{2}$$

where $dh/dt|_{initial}$ is the initial linear foam decay rate.

Each experimental test on foamability and foam decay was processed, thus obtaining the Bickerman unit of foaminess, the initial speed of froth decay, and the foam production. It will be shown here what these three parameters look like for one of the cases and after that, it will be operated only with foam production as a basic result of this study.

3. Results and Discussion

Frothers are used to facilitate the dispersion of air into fine-sized bubbles and to stabilize the foam. And, as reported in our previous study that the strength of frothers can be determined by either the dynamic foamability index (DFI) or their critical coalescence concentration values [17,20].

For this aim, a dynamic foam analyzer (DFA) is used to evaluate the foamability and the rate of foam decay. The foam height and lifetime of a bubble are quantitative parameters that represent dynamic stability. In the DFA technique, the foam rises in a column, and the height is measured as a function of time. The maximum height can be well correlated with foamability [21,22]. The flotation process used in ore preparation depends on the formation of a carefully controlled and stable foam. It also controls the amount of water recovered and thus the amount of mechanical transport. Measuring the foam height for a given frother dosage at a given airflow rate in a metering cylinder is a simple, semi-quantitative method of estimating the frothing power of a frother. Frothers adsorb at the gas-liquid interface and change the interface properties. Surfactants play an important role in foam stability. The effect of surfactants on foam behavior is not clear. In flotation, the foam must be dynamically stable as the foam is formed, that is the foam must be stable to bubble coalescence. However, Surfactants play an important role in foam stability. Unstability is undesirable in processes such as pumping or filtration. But it should be noted that both foamability and foam decay are mainly controlled by the particles rather than the surfactant molecules [23,24].

3.1. Dynamic Foam Stability Experiments

3.1.1. Foam Stability and Foam Decaying of Frothers

First, the differences in foamability and foam decay at different concentrations of frothers were investigated. In Figure 3, the results of foamability and foam decaying are shown separately in the critical coalescence concentrations (CCC) values for each frother. The values for these frothers were both experimentally determined and theoretically calculated in our previous studies [25,26].

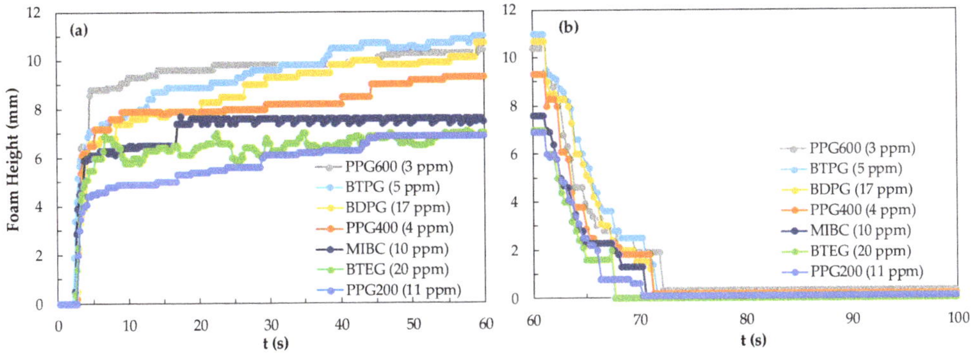

Figure 3. (**a**) Foamability and (**b**) foam decay of frothers at their CCC values.

A unique relationship was found between foam decay (FD), Mw (Molecular weight), and foamability of the homologous foaming series. However, when the functional groups of the molecule are changed, different results were obtained. Therefore, seven foams with specific values of CCC were analyzed. According to the CCC values of seven foams obtained from this study, foaming agents should be considered as PPG, BTPG, BDPG, MIBC, and BTEG depending on the homologous series with increasing carbon atoms, taking into account the type and location of functional groups. As seen in Figure 3, the highest foamability measured as 11 mm and lowest foam decay measured as 6.8 mm were obtained with PPG600 at CCC values, while the lowest foamability and foam decaying values measured as 0 mm were obtained with PPG200.

In Figure 4, the foamability and foam decay of PPG600 frother from 1 ppm to 1000 ppm were examined. The results of this series of tests showed that no significant difference could be obtained for concentrations over 10 ppm. Although foam height at the CCC value of 3 ppm at 60 s is 10 mm, the foam decay at and below 3 ppm is rather sharp indicating the need for higher frother concentrations to maintain relatively acceptable foam stabilities.

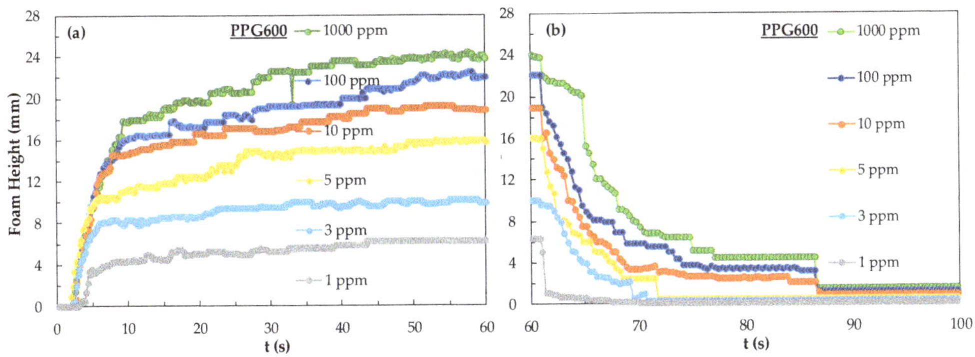

Figure 4. (**a**) Foamability and (**b**) foam decay of PPG600 frother at different concentrations.

3.1.2. Foam Stability and Foam Decaying of Collectors

As it is known, foamability and foam decay effect are two of the most important classification parameters for determining foam properties. Interestingly excluding sulfide minerals and in some cases coal flotation systems, all other minerals require some kind of frother-collector combinations. Therefore, the effect of collector concentration should be considered to comment on foamability and foam decay properties, and, accordingly, the flotation of fine-sized particles. In addition to these parameters, the bubble sizes in terms of Sauter Mean Diameter (SMD) values and the bubble size distribution (BSD) under the control of different parameters such as gas flow rate, the pore size of the porous frit may also be effective on tuning the flotation conditions of fine particles [27]. Thus, in this study, the characteristics of different frothers and frother + collector mixtures were studied in order to provide a different perspective on the aforementioned issue of the flotation of very fine particles. Here, both the difference between the collectors and the foamability and foam decay measurements were carried out at different concentrations of the collectors. As seen in Figure 5, the foamability was found to be more durable and foam decay time increased as the concentration of the DAH collector increased. As seen in Figure 5, foaming begins at a DAH concentration of 5.10^{-5} mol/L and increases, more above 1.10^{-4} mol/L. It seems that the foam decaying continues for 80 s at the concentration of 7.10^{-4} mol/L DAH and 1.10^{-3} mol/L DAH.

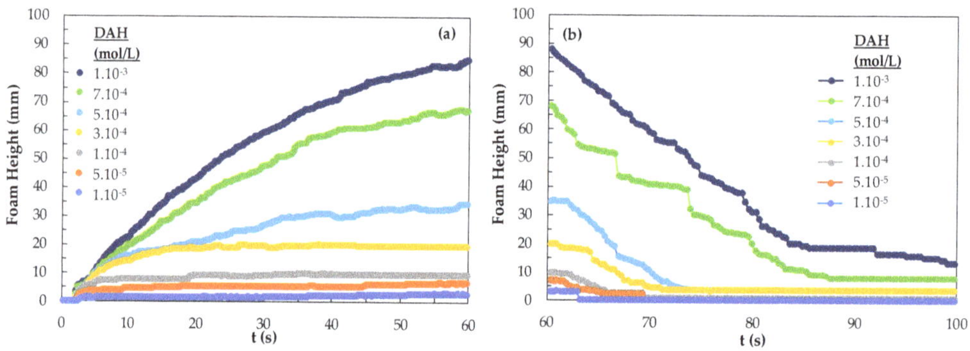

Figure 5. (**a**) Foamability and (**b**) foam decay of DAH collector as a function concentration.

The results of foamability and foam decay of NaOL are given in Figure 6. In Figure 6, foam starts to rise at the 5.10^{-5} mol/L concentration of NaOL and shows the highest foaming at 5.10^{-4} mol/L. Foam decay appears to be 18 mm after 2 h 25 min for 5.10^{-4} mol/L (inset figure). NaOL 1.10^{-3} mol/L has the highest stability and the foam decaying was the lowest. Also, too-high foam stability is not suitable for flotation and was considered too high a concentration. The effective concentration range of NaOL on foam behavior is much different from that of the DAH collector. That is NaOL has 18 carbon atoms, and 33 hydrogen atoms, the molecular mass is 304.4 g/mol, and the functional group is carboxyl, while DAH has 12 carbon atoms, 28 hydrogen atoms, the molecular mass is 221.8, and one amino functional group. The low number of hydrogens in the chain greatly affects the surface and the number and location of functional groups. Because the chemistry adsorption of NaOL is slower, NaOL increases the potential for hydrogen formation. Infrared spectrophotometry studies showed that the RCOO− ion was chemically adsorbed on the magnesite surface and physically adsorbed on the serpentine surface, while the NH_3^+ ion was physically adsorbed on the surface of both minerals [28]. The higher the concentration of the DAH collector, the higher the foam whereas the foam rise between 5.10^{-5}–5.10^{-4} mol/L is sufficient for NaOL. This difference is related to the molecular weight, content, and molecular structure of the collectors.

It was also found that physical adsorption had a medium impact on foamability and foam decay while chemical adsorption maintained better foamability and foam decay. Thus, the chemical adsorption of NaOl with magnesite particles and the physical adsorption characteristics of DAH with quartz are shown in Figure 7.

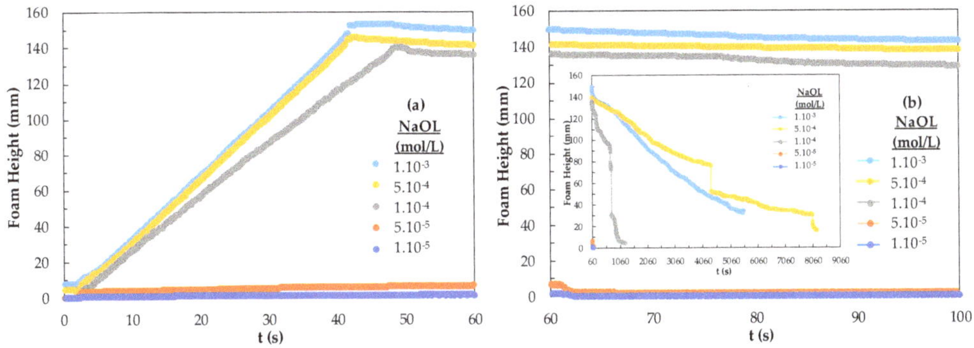

Figure 6. (**a**) Foamability and (**b**) foam decay according to the concentration of the NaOL collector.

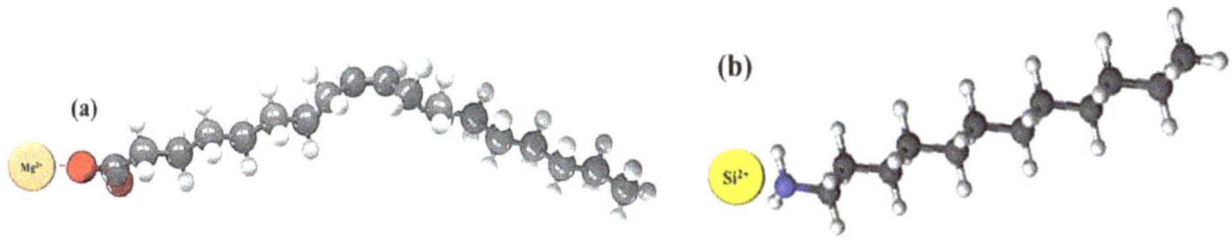

Figure 7. (**a**) chemistry adsorption of magnesite with NaOL and (**b**) physical adsorption of quark with DAH.

3.1.3. Frothers in the Presence of Collectors

Continuous drainage for enrichment depends on the stability of the foam produced in the flotation cell. Therefore, foam structure and stability are very important in the flotation process. To achieve adequate separation in flotation, additional reagents that act as frothers, collectors, or both must be used [29,30]. The type and amount of reagent added are important parameters for flotation. On the other hand, the type of the collector in particular also significantly affects the characteristics of the frother in the mixture. In a recent study [31], it was found that the use of DAH as a component of a frother + collector mixture resulted in lowering the CCC values of frothers such as PPG200, 400, and 600 due to its additional frother property during flotation of particles. This in turn will result in finer bubble sizes which will affect the flotation recoveries of very fine particles. Taking into account that knowledge in mind, in this study, the effects of mixtures of collectors and frothers on foamability and foam decay were investigated. In the first experiment, the effects of pure reagents on foamability and foam decay performance were determined. The effects of reagent type and dosage on foamability and foam decay were investigated in foam flotation. This study aims to explain the relationship between foamability, foam decay, and mixing system depending on the effect of the reagent. The results of foam stability and foam degradation measurements for frothers (CCC values) as a function of DAH appear in Figure 8.

As seen in Figure 8, larger foamability is achieved with PPG600 + DAH compared to other frothers whereas the least foamability was observed in the presence of BTEG + DAH. When 5.10^{-5} mol/L DAH was mixed with a frother, the foam formed with 3 ppm PPG600 mixture increased up to 48.8 mm within 60 s and the foam degradation decreased to 4.3 mm after 70 s.

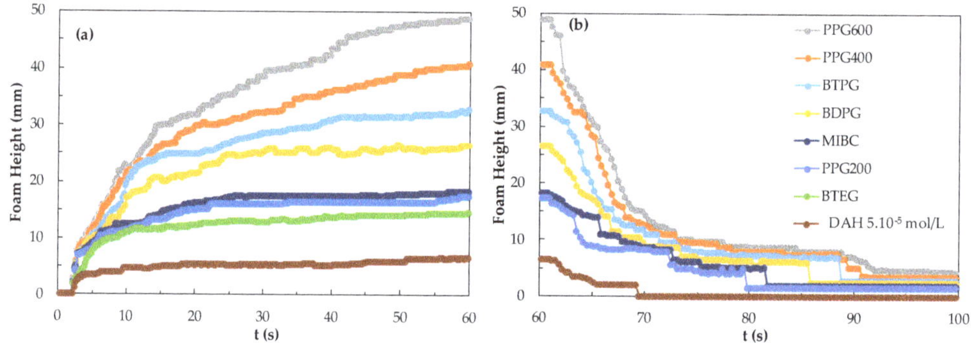

Figure 8. (**a**) Foamability and (**b**) foam decay for the frothers at their CCC values in the presence of 5.10^{-5} mol/L DAH.

The results of foamability and foam degradation measurements by mixing 5.10^{-5} mol/L NaOL with the frothers at CCC concentrations are shown in Figure 9. As can be seen from Figure 9, the CCC values of the frothers in the presence of 5.10^{-5} mol/L NaOL show the highest foamability and lowest foam decay in the PPG600 mix system, while BTEG shows the lowest foam stability and highest foam decay. Thus, from this, it is determined that NaOL and PPG600 are more suitable for using a mixing system.

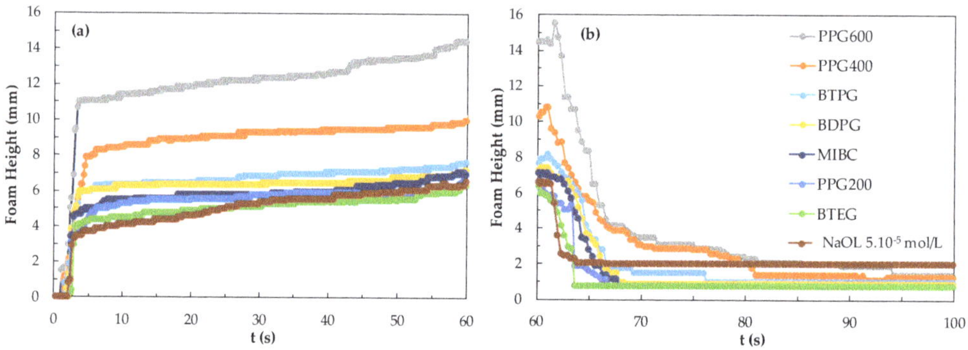

Figure 9. (**a**) Foamability and (**b**) foam decay for the frothers at their CCC values in the presence of 5.10^{-5} mol/L NaOL.

Also, the mixed system with 5.10^{-5} mol/L DAH and frothers showed better foamability than the mixed system with 5.10^{-5} mol/L NaOL and frothers; this probably depends on the length of the hydrocarbon and the adsorption kinetics. This is because longer hydrocarbon chains not only adsorb more slowly but also produce more stable adsorption. The molecular mass and chain length of DAH are lower than that of NaOL, which means better interaction with foaming agents and better foamability. Concentration is shown to be important when mixing a strong collector with a strong foamer. Therefore, the foamability and foam decay of frother and collector mixture with 3 ppm of PPG600 at different concentrations of collectors were investigated.

Figure 10 shows the foamability and foam degradation results for the DAH header in the presence of 3 ppm PPG600 frother. Here, the highest foamability and lowest foam degradation of 5.10^{-5} mol/L DAH are given, while the lowest foamability of 1.10^{-3} mol/L M DAH is shown, according to the mixing system of PPG600 3 ppm and DAH collector concentrations. In the presence of 3 ppm PPG600, the foamability of the DAH mix system decreased to 5.10^{-5} mol/L > 1.10^{-4} mol/L > 1.10^{-5} mol/L > 5.10^{-4} mol/L >1.10^{-3} mol/L, and the foam decay was 1.10^{-4} mol/L > 5.10^{-4} mol/L > 1.10^{-3} mol/L > 5.10^{-5} mol/L >

1.10^{-5} mol/L. In the presence of a strong frother PPG600 and strong collector DAH mix system, low concentration seems to be suitable. It also means that it is more suitable to use a mixing system with DAH collector + PPG600 (3 ppm) frother.

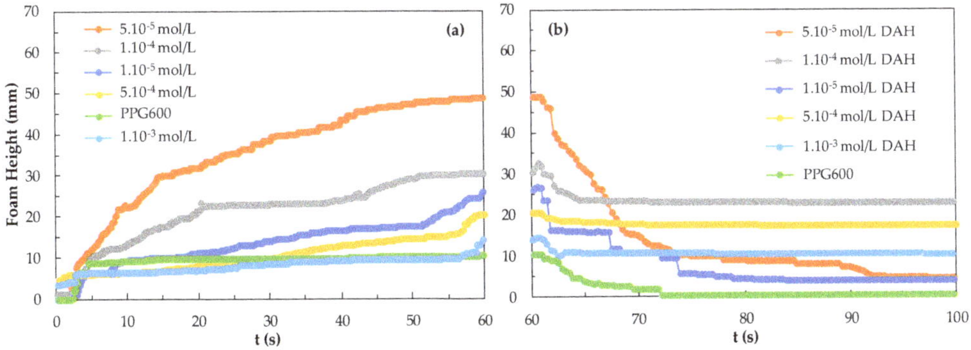

Figure 10. (**a**) Foamability and (**b**) foam decay results for the DAH collector in the presence of 3 ppm PPG600 frother.

Figure 11 shows the foamability and foam decay of the mixed system with the concentrations of NaOL in the presence of 3 ppm PPG600. The highest foamability and foam decay in the mixed system of 3 ppm PPG600 and 5.10^{-4} mol/L and 1.10^{-3} mol/L NaOL, the least foamability and foam decay were observed in the 1.10^{-5} M and 5.10^{-5} mol/L NaOL with 3 ppm PPG600 system. Therefore, it shoractically should be understood that pure NaOL showed to act both as a collector and frother.

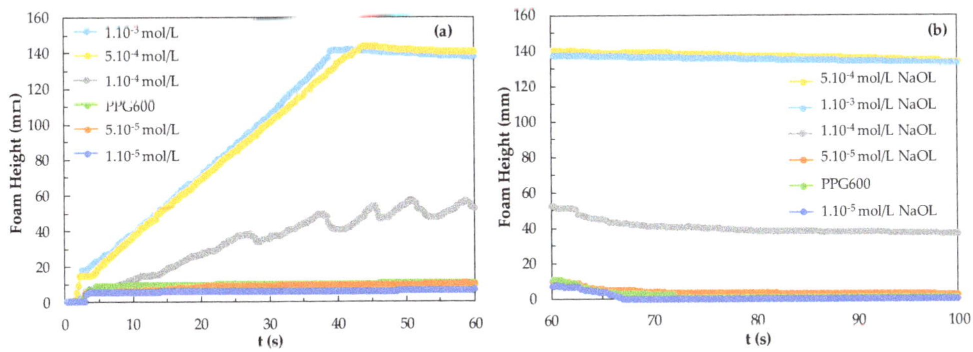

Figure 11. (**a**) Foamability and (**b**) foam decay results for NaOL collector in the presence of 3 ppm PPG600.

3.2. Modeling Studies

Figure 12 shows the foaminess, the initial speed of froth decay, and the foam production of PPG600 versus its concentration. One can be seen in Figure 12 that as the foaminess increases, the rate of foam decay generally decreases, and the foam production increases with the increase of the concentration of PPG600. Yet, it is interesting the way, in which these variations of the parameters occur because foam production is the line between foamability and foam decay. These two parameters are usually competing, thus confusing the appropriate reading of the experimental data.

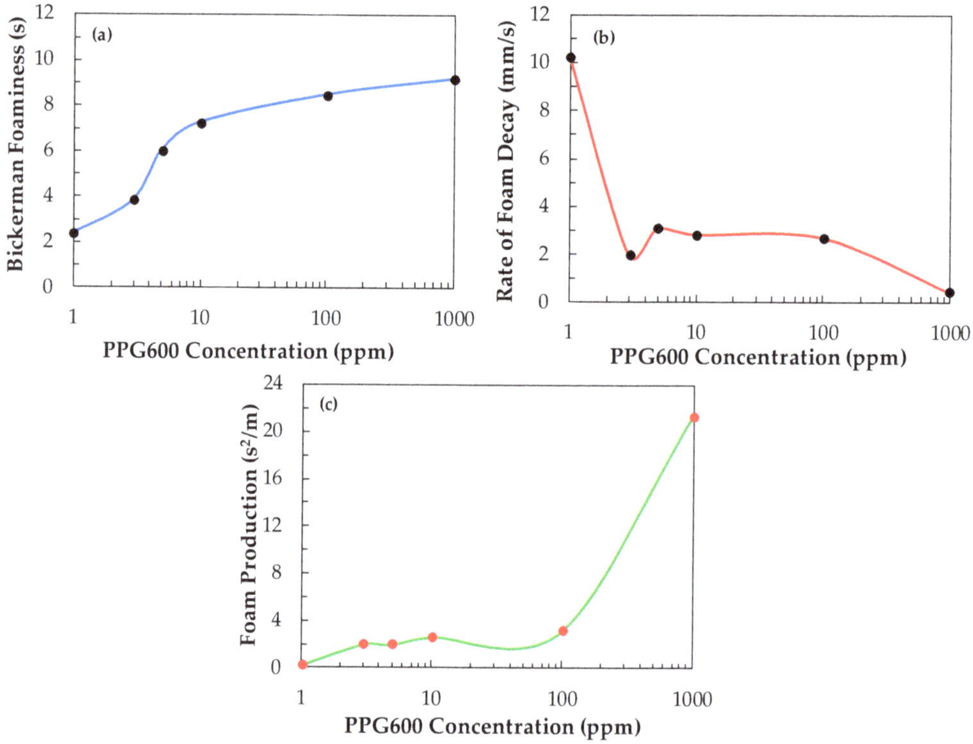

Figure 12. (**a**) Bickerman foaminess, (**b**) rate of foam decay, and (**c**) foam production vs. the concentration of PPG600.

Figure 13 shows the foam production (FP) [19] of the basic frothers, studied in this work, at their CCC values in the absence and presence of collectors (DAH and NaOL). Also, in the absence of collectors PPG200, PPG400, and PPG600, BTEG, and MIBC have almost the same FP values, while BDPG and BTPG produce larger FP values. It is interesting to know how the FP values of the frothers change upon adding 5.10^{-5} mol/L collectors. The latter concentration of the collectors is operational in the flotation process. For example, the addition of DAH to PPG200, PPG400, and PPG600 increase significantly their FP values, The FP value of BDPG is not affected while the FP values of BTPG and BTEG are decreased, while this one of MIBC is increased. Upon the addition of NaOL, the increase of the FP values of PPG200, PPG400, and PPG600 is weaker. The FP values of BDPG and BTPG are decreased. This corresponds to lower stationary foam height and/or faster foam decay and indicates the quality of froth according to analysis. The frothers of BTEG and MIBC are not practically affected by the presence of NaOL. This information can be exploited in the real frother/collectors formulations to decrease the level of entrainment of particles. Low foam production corresponds to faster-decaying foam, thus releasing the water on top of the foam containing entrained particles. The collectors (NaOL and DAH) have certain surface-active properties. Therefore, their froth performance was studied separately.

Figure 14 shows how the FP values of NaOL and DAH generally increase with increasing their concentration. For example, NaOL has small FP values until C = 5.10^{-5} mol/L, above which, its FP value increases substantially, i.e., it converts into a strong frother. The FP values of DAH increase with its concentration in the way given in Figure 13. As far as PPG 600 is the best-recognized frother, producing fine bubbles in the pulp zone, it was focused on its performance in presence of these two frothers. The interaction between frother and collector, which is difficult to predict, will result in their mutual froth performance.

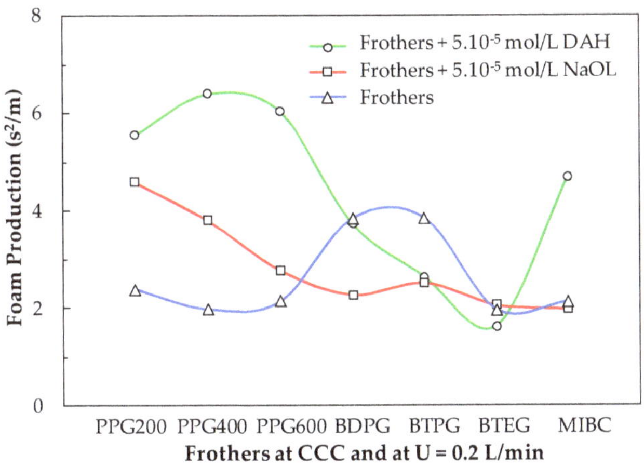

Figure 13. Foam production of the basic frothers, studied in this work, with concentrations equal to their CCC values in the absence and presence of 5.10^{-5} mol/L collectors (DAH and NaOL).

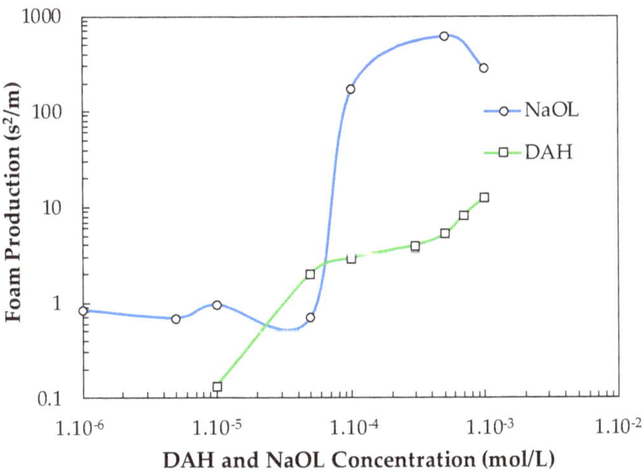

Figure 14. Foam production of NaOL and DAH versus their concentration.

Figure 15 shows the FP values of NaOL and NaOL + 3 ppm PPG600 vs. NaOL concentration. It is seen from Figure 15 that the addition of 3 ppm PPG600 has an additive effect on the total FP value until the critical concentration C = 5.10^{-5} mol/L, above which interaction between them occurs, thus resulting in a smaller total FP value compared with that of the single collector.

The FP values of DAH and DAH + 3 ppm PPG600 vs. the concentration of DAH are shown in Figure 16. One can see that the contribution of PPG600 to this system is practically additive.

The frothing performance of a mixture of frother and collector in terms of foam production could be used for the proper selection of reagent formulation at industrial conditions. Of course, it depends on the operational conditions of each plant, but this approach can be used for every specific industrial case. For example, the presence of coarse bubbles in the froth zone minimizes the entrainment of particles. This corresponds to small foam production (FP) value. Therefore, it was interesting to choose such a combination of frother and collector, which results in as small as possible FP value. For example, the combination of 5.10^{-5} mol/L DAH and 3 ppm PPG600 have a small FP value and can be used for obtaining a froth zone with coarse bubbles and a minimal level of entrainment. According to Figure 15, 5.10^{-5} mol/L NaOL + 3 ppm PPG600 is the most suitable combination having a small FP value. One should not use a concentration of NaOL larger than 5.10^{-5} mol/L because its FP values significantly increased.

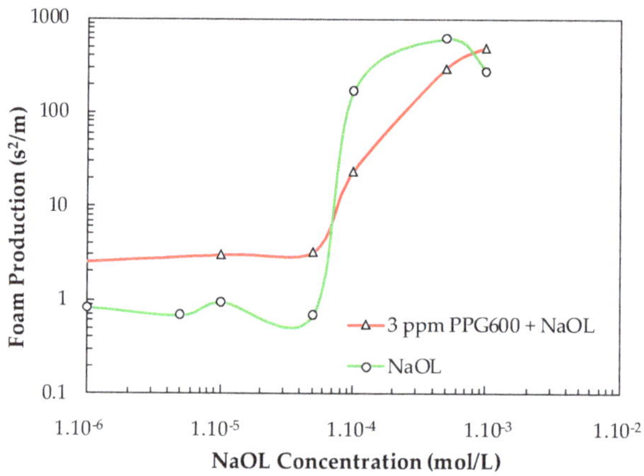

Figure 15. Foam production value of NaOL and NaOL + 3 ppm PPG600 vs. NaOL concentration.

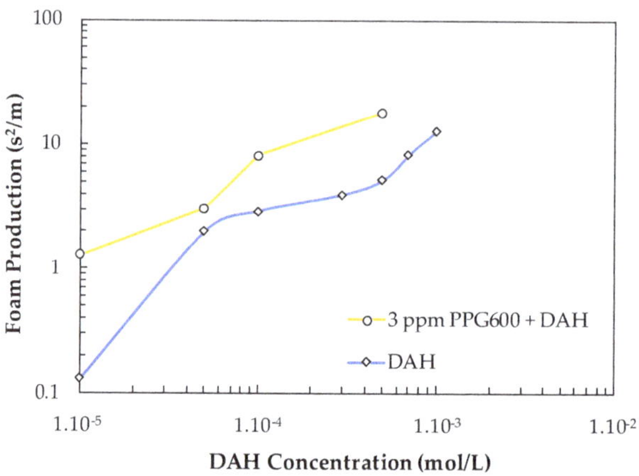

Figure 16. Foam production value of DAH and DAH + 3 ppm PPG600 vs. DAH concentration.

4. Conclusions

The results of this study revealed that foamability and foam decay significantly vary with the kind of reagent. In the presence of frother, while the highest foamability and low foam decaying were obtained with PPG600 at its CCC values, the lowest foamability, and high foam decay were found in the presence of BTEG. Among all the frothers, foam stability, and foam degradation were exhibited in the order BTEG < PPG200 ≈ MIBC < BDPG < BTPG < PPG400 < PPG600.

In the presence of sodium oleate (NaOL) and DAH collectors, foam decay began at a concentration of 5.10^{-5} mol/L. It was observed that as the concentration of the collectors increased, the foamability was larger and the foam decay time increased. The highest foam formation and the highest foam height were found to be ~98 mm at 1.10^{-3} mol/L for DAH and 152 mm at 5.10^{-4} mol/L for NaOL. The foam heights obtained with NaOL reached decayed in 40 s at the shortest time. NaOL and DAH collectors are indicated to support frothers in foam formation. It was determined that when the concentration of NaOL increased in the presence of 3 ppm PPG600, the foam height increased further, and the foamability was larger. This was found to be supported by NaOL and DAH collectors in the presence of 3 ppm PPG600 frother. It is shown that it is more appropriate to use a combination of 3 ppm PPG600 frother with DAH or NaOL collectors.

Overall, it can be concluded from this study that the foam production concept can be used for the proper selection of frothers and collectors producing coarse bubbles with small entrainment in the froth zone. And, the mixture of frother and collector results in either: (i) additive effect or (ii) interaction between them, which can either decrease or increase the FP value of the mixture.

The preliminary theoretical prediction of the frothing performance of frother + collector is a quite complex subject, but its experimental study, shown in this work, could help us to better understand the absence or presence of synergy between them.

Author Contributions: Experimental methodology, O.G., K.B., O.O., F.B.; validation, S.I.K., N.A.G.; investigation, O.G., K.B., O.O.; resources, O.G., O.O., F.B., M.S.Ç.; original draft preparation, O.G. and O.O.; writing—review and editing, O.G., O.O., S.I.K., M.S.Ç.; supervision, O.O., M.S.Ç.; project administration, M.S.Ç. All authors have read and agreed to the published version of the manuscript.

Funding: This paper is supported by European Union's Horizon 2020 research and innovation program under grant agreement No. 821265, project FineFuture (Innovative technologies and concepts for fine particle flotation: unlocking future fine-grained deposits and Critical Raw Materials resources for the EU).

Data Availability Statement: Not applicable.

Conflicts of Interest: The authors declare no conflict of interest.

Appendix A

The following data was obtained from the results of modeling studies:

Table A1. CCC values, Bikerman unit of foaminess, the initial rate of foam decay, foam production, and the standard deviation of the foam production value of selected frothers.

Frothers	CCC (ppm)	Foaminess (s)	dh/dt (mm/s)	Foam Production (s^2/mm)	St. Dev. (s^2/mm)
PPG200	11	2.60	1.09	2.38	0.18
PPG400	4	3.51	1.78	1.97	0.11
PPG600	3	3.92	1.84	2.13	0.11
BDPG	17	4.03	1.05	3.84	0.19
BTPG	5	4.15	1.08	3.85	0.19
BTEG	20	2.68	1.37	1.96	0.15
MIBC	10	2.90	1.37	2.12	0.15

Table A2. Bikerman unit of foaminess, the initial rate of foam decay, foam production, and the standard deviation of the foam production value for different concentrations of PPG600.

	C (ppm)	Foaminess (s)	dh/dt (mm/s)	Foam Production (s^2/mm)	St. Dev. (s^2/mm)
PPG600	1	2.38	10.21	0.23	0.02
	3	3.85	1.95	1.97	0.10
	5	6.03	3.05	1.98	0.07
	10	7.28	2.79	2.61	0.07
	100	8.48	2.67	3.18	0.04
	1000	9.20	0.43	21.38	0.23

Table A3. Bikerman unit of foaminess, the initial rate of foam decay, foam production, and the standard deviation of the foam production value for different concentrations of DAH.

	C (mol/L)	Foaminess (s)	dh/dt (mm/s)	Foam Production (s^2/mm)	St. Dev. (s^2/mm)
DAH	1.10^{-5}	1.17	8.82	0.13	0.01
	5.10^{-5}	2.49	1.23	2.02	0.08
	1.10^{-4}	3.69	1.26	2.93	0.08
	3.10^{-4}	7.65	1.93	3.96	0.05
	5.10^{-4}	13.12	2.53	5.19	0.04
	7.10^{-4}	27.37	3.31	8.26	0.03
	1.10^{-3}	32.35	2.52	12.82	0.04

Table A4. CCC values, Bikerman unit of foaminess, the initial rate of foam decay, foam production, and the standard deviation of the foam production value of the frothers at 5.10^{-5} mol/L DAH.

	CCC (ppm)	Foaminess (s)	dh/dt (mm/s)	Foam Production (s^2/mm)	St. Dev. (s^2/mm)
PPG200	11	6.56	1.18	5.54	0.08
PPG400	4	15.42	2.41	6.39	0.04
PPG600	3	18.40	3.05	6.03	0.03
BDPG	17	10.03	2.70	3.72	0.04
BTPG	5	12.37	4.70	2.63	0.02
BTEG	20	5.50	3.40	1.62	0.03
MIBC	10	6.90	1.47	4.69	0.07

Table A5. CCC values, Bikerman unit of foaminess, the initial rate of foam decay, foam production, and the standard deviation of the foam production value of the frothers at 5.10^{-5} mol/L NaOL.

	CCC (ppm)	Foaminess (s)	dh/dt (mm/s)	Foam Production (s^2/mm)	St. Dev. (s^2/mm)
PPG200	11	2.49	0.54	4.59	0.18
PPG400	4	4.07	1.07	3.79	0.09
PPG600	3	5.88	2.13	2.76	0.05
BDPG	17	2.79	1.23	2.27	0.08
BTPG	5	3.09	1.23	2.51	0.08
BTEG	20	2.41	1.17	2.06	0.09
MIBC	10	2.68	1.36	1.97	0.07

Table A6. Bikerman unit of foaminess, the initial rate of foam decay, foam production, and the standard deviation of the foam production value for different concentrations of DAH in presence of 3 ppm PPG600.

	C (mol/L)	Foaminess (s)	dh/dt (mm/s)	Foam Production (s^2/mm)	St. Dev. (s^2/mm)
DAH	0	3.92	1.84	2.13	0.11
	1.10^{-5}	11.65	8.94	1.30	0.01
	5.10^{-5}	20.06	6.53	3.07	0.02
	1.10^{-4}	20.85	2.53	8.22	0.04
	5.10^{-4}	14.14	0.78	18.12	0.13

Table A7. Bikerman unit of foaminess, the initial rate of foam decay, foam production, and the standard deviation of the foam production value for different concentrations of NaOL in presence of 3 ppm PPG600.

	C (mol/L)	Foaminess (s)	dh/dt (mm/s)	Foam Production (s^2/mm)	St. Dev. (s^2/mm)
NaOL	0	3.92	1.84	2.13	0.11
	1.10^{-5}	2.83	0.94	3.00	0.11
	5.10^{-5}	4.60	1.43	3.22	0.07
	1.10^{-4}	35.32	1.50	23.51	0.07
	5.10^{-4}	104.54	0.35	297.24	0.28
	1.10^{-3}	103.52	0.21	498.42	0.48

References

1. George, P.; Nguyen, A.V.; Jameson, G.J. Assessment of true flotation and entrainment in the flotation of submicron particles by fine bubbles. *Miner. Eng.* **2004**, *17*, 847–853. [CrossRef]
2. Benzaazoua, M.; Bussiere, B.; Kongolo, M.; McLaughlin, J.; Marion, P. Environmental desulphurization of four Canadian mine tailings using froth flotation. *Int. J. Miner. Process.* **2000**, *5*, 57–74. [CrossRef]
3. Mbamba, C.K.; Harrison, S.T.L.; Franzidis, J.P.; Broadhurst, J.L. Mitigating acid rock drainage risks while recovering low-sulfur coal from ultrafine colliery wastes using froth flotation. *Miner. Eng.* **2012**, *29*, 13–21. [CrossRef]
4. Yoon, R.H. The role of hydrodynamic and surface forces in bubble–particle interaction. *Int. J. Miner. Process.* **2000**, *58*, 129–143. [CrossRef]

5. Karakashev, S.I.; Grozev, N.A.; Ozdemir, O.; Guven, O.; Ata, S.; Bournival, G.; Batjargal, K.; Boylu, F.; Hristova, S.; Çelik, M.S. Physical restrictions of the flotation of fine particles and ways to overcome them. *Physicochem. Probl. Miner. Process.* **2022**, *58*, 153944. [CrossRef]
6. Severov, V.V.; Filippova, I.V.; Filippov, L.O. Use of fatty acids with an ethoxylated alcohol for apatite flotation from old fine-grained tailings. *Miner. Eng.* **2022**, *188*, 107832. [CrossRef]
7. Foucaud, Y.; Collet, A.; Filippova, I.V.; Badawi, M.; Filippov, L.O. Synergistic effects between fatty acids and non-ionic reagents for the selective flotation of scheelite from a complex tungsten skarn ore. *Miner. Eng.* **2022**, *182*, 107566. [CrossRef]
8. Derjaguin, B.V.; Dukhin, S.S.; Rulyov, N.N. Kinetic Theory of Flotation of Small Particles. In *Surface and Colloid Science*; Matijević, E., Good, R.J., Eds.; Springer: Boston, MA, USA, 1984.
9. Pyke, B.; Fornasiero, D.; Ralston, J. Bubble particle heterocoagulation under turbulent conditions. *J. Colloid Interface Sci.* **2003**, *265*, 141–151. [CrossRef]
10. Miettinen, T.; Ralston, J.; Fornasiero, D. The limits of fine particle flotation. *Miner. Eng.* **2010**, *23*, 420–437. [CrossRef]
11. Abrahamson, J. Collision rates of small particles in a vigorously turbulent fluid. *Chem. Eng. Sci.* **1975**, *30*, 1371–1379. [CrossRef]
12. Schubert, H. On the turbulence-controlled microprocesses in flotation machines. *Int. J. Miner. Process.* **1999**, *56*, 257–276. [CrossRef]
13. Gaudin, A.M.; Groh, J.O.; Henderson, H.B. *Effect of Particle Size on Flotation*; AIME Technical Publications: Englewood, NJ, USA, 1931; Volume 414, pp. 3–23.
14. Yalcin, T.; Byers, A. Dissolved gas flotation in mineral processing. *Miner. Process. Extr. Metall. Rev.* **2006**, *27*, 87–97. [CrossRef]
15. Tao, Y.J.; Liu, J.T.; Yu, S.; Tao, D. Picobubble enhanced fine coal flotation. *Sep. Sci. Technol.* **2006**, *41*, 3597–3607. [CrossRef]
16. Geldenhuys, S.; McFadzean, B. Column diameter effects on dynamic froth stability measurement. In Proceedings of the XXIX International Mineral Processing Congress (IMPC 2018), Moscow, Russia, 17–21 September 2018.
17. Batjargal, K.; Guven, O.; Ozdemir, O.; Boylu, F.; Çelik, M.S. Effect of frother and collector mixture on froth stability in ultra-fine size quartz/DAH flotation system. In Proceedings of the 17th International Mineral Processing Symposium (IMPS 2022), Istanbul, Turkiye, 15–17 December 2022.
18. Bikerman, J.J. The unit of foaminess. *Trans. Faraday Soc.* **1938**, *34*, 0634–0638. [CrossRef]
19. Karakashev, S.I.; Georgiev, P.; Balashev, K. Foam production—Ratio between foaminess and rate of foam decay. *J. Colloid Interface Sci.* **2012**, *379*, 144–147. [CrossRef]
20. Karakashev, S.I.; Grozev, N.A.; Ozdemir, O.; Batjargal, K.; Guven, O.; Ata, S.; Bournival, G.; Boylu, F.; Çelik, M.S. On the frother's strength and its performance. *Miner. Eng.* **2021**, *171*, 107093. [CrossRef]
21. Pugh, R.J. *Bubble and Foam Chemistry*; CUP, Cambridge Press: Cambridge, UK, 2016; Volume 12, pp. 405–419.
22. Schwarz, S. The Relationship between Froth Recovery and Froth Structure. Ph.D. Thesis, Ian Wark Research Institute, University of South Australia, Adelaide, Australia, 2004.
23. Neethling, S.J.; Brito-Parada, P.R. Predicting flotation behaviour the interaction between froth stability and performance. *Miner. Eng.* **2018**, *120*, 60–65. [CrossRef]
24. Holmberg, K. *Surfactant and Polymers in Aqueous Solution*, 2nd ed.; John Wiley & Sons Ltd.: Chichester, UK, 2003.
25. Guven, O.; Batjargal, K.; Ozdemir, O.; Karakashev, S.I.; Grozev, N.A.; Boylu, F.; Çelik, M.S. Experimental procedure for the determination of the critical coalescence concentration (CCC) of simple frothers. *Minerals* **2020**, *10*, 617. [CrossRef]
26. Karakashev, S.I.; Grozev, N.A.; Batjargal, K.; Guven, O.; Özdemir, O.; Boylu, F.; Çelik, M.S. Correlations for easy calculation of the critical coalescence concentration (CCC) of simple frothers. *Coatings* **2020**, *10*, 612. [CrossRef]
27. Batjargal, K.; Guven, O.; Ozdemir, O.; Karakashev, S.I.; Grozev, N.A.; Boylu, F.; Çelik, M.S. Adsorption Kinetics of Various Frothers on Rising Bubbles of Different Sizes under Flotation Conditions. *Minerals* **2021**, *11*, 304. [CrossRef]
28. Gence, N.; Ozdag, H. Surface properties of magnesite and surfactant adsorption mechanism. *Int. J. Miner. Process.* **1995**, *43*, 37–47. [CrossRef]
29. Woodburn, E.T.; Flynn, S.A.; Cressey, B.A.; Cressey, G. The effect of froth stability on the beneficiation of low-rank coal by flotation. *Powder Technol.* **1984**, *40*, 167–177. [CrossRef]
30. Subrahmanyam, T.V.; Forssberg, E. Froth Stability Particle Entrainment and Drainage in Flotation. *Int. J. Miner. Process.* **1988**, *23*, 33–53. [CrossRef]
31. Batjargal, K.; Guven, O.; Ozdemir, O.; Karakashev, S.; Grozev, N.A.; Boylu, F.; Çelik, M.S. Bubbling properties of frothers and collectors mix system. *Physicochem. Probl. Miner. Process.* **2022**, *58*, 152890. [CrossRef]

minerals

MDPI

Article

Effect of Nanobubbles on the Flotation Behavior of Microfine-Grained Serpentine

Bingang Lu [1,2], Weiguang Xu [3,4,5,6,7,*], Chunhua Luo [1,2,*], Wenjuan Li [3,4,5,6,7], Xiaohui Su [1,2], Yongsheng Song [3,4,5,6,7], Jianhang Zhou [3,4,5,6,7] and Kaiguo Li [3,4,5,6,7]

1 National Key Laboratory of Ni&Co Associated Minerals Resources Development and Comprehensive Utilization, Jinchang 737100, China; 13830587684@163.com (B.L.); xiaohui8001@163.com (X.S.)
2 Jinchuan Nickel & Cobalt Research and Engineering Insitute, Jinchang 737100, China
3 National Engineering Research Center for Environment-Friendly Metallurgy in Producing Premium Non-Ferrous Metals, GRINM Group Corporation Limited, Beijing 100088, China; juanzi88888@126.com (W.L.); sysmba@163.com (Y.S.); zhoujianhang8888@163.com (J.Z.); likaiguo139@139.com (K.L.)
4 GRINM Resources and Environment Tech. Co., Ltd., Beijing 100088, China
5 General Research Institute for Nonferrous Metals, Beijing 100088, China
6 Beijing Engineering Research Center of Strategic Nonferrous Metals Green Manufacturing Technology, Beijing 100088, China
7 GRIMAT Engineering Institute Co., Ltd., Beijing 101407, China
* Correspondence: xuweiguang@grinm.com (W.X.); china540@163.com (C.L.); Tel.: +86-10-18713827621 (W.X.); +86-10-15097020780 (C.L.)

Abstract: At present, scholars mainly study the relationship between nanobubbles and useful minerals, often ignoring the influence of bubbles on fine gangue minerals. When selecting nickel sulfide ore, scholars often faced with mudded and irrepressible serpentine, which seriously affects the quality of the concentrate. This mudded serpentine mineral often enters foam products with bubbles. In this study, the role of nanobubbles in the flotation behavior of hydrophilic serpentine was examined. Nanobubbles were successfully prepared via ultrasonic cavitation, with sizes ranging from 50 to 250 nm. The size and number of bubbles produced at 1 min and 2 min of sonication were significantly better than those of the prolonged test group, and it was found that longer sonication time did not produce better results. The stability of the nanobubbles produced via ultrasound was studied, and it was found that the nanobubbles were stable, with no change in size and only a slight decrease in number as the resting time increased. Nanobubbles were introduced into serpentine flotation, we found that the presence of nanobubbles significantly reduced the flotation recovery of serpentine. The presence of nanobubbles reduced the froth entrainment rate of microfine-grained serpentine, which in turn reduced its flotation rate. In the depressant group trials, it was found that the nanobubbles also reduced the amount of depressant. In short, the presence of nanobubbles can prevent the floating of fine hydrophilic gangues during flotation.

Keywords: nanobubbles; serpentine; bubble diameter; recovery; entrainment

check for updates

Citation: Lu, B.; Xu, W.; Luo, C.; Li, W.; Su, X.; Song, Y.; Zhou, J.; Li, K. Effect of Nanobubbles on the Flotation Behavior of Microfine-Grained Serpentine. *Minerals* **2023**, *13*, 1299. https://doi.org/10.3390/min13101299

Academic Editor: Jan Zawala

Received: 14 August 2023
Revised: 20 September 2023
Accepted: 2 October 2023
Published: 7 October 2023

1. Introduction

Froth flotation represents a widely employed technique within the realm of mineral in mineral processing, particularly in the context of sulfide ore separation. It is a physicochemical process based on assessing the surface chemistry differences in the presence of flotation chemicals between valuable/economical minerals and gangue minerals using bubbles in the pulp. The particle size of minerals is an essential factor affecting flotation efficiency [1]. For the conventional flotation of sulfide ores, the optimal particle size range typically falls within the spectrum of 0.01 to 0.3 mm. Still, with the depletion of high-quality resources, it becomes necessary to process poor ores with extremely fine inlay grain size, which is somewhat overwhelming for conventional flotation techniques. In this regard, numerous studies have found that nanobubble flotation can improve the flotation of fine-grained minerals [2,3]. Nanobubbles, alternatively referred to as microbubbles, are minute bubbles with diameters measuring less than 1 μm [4]. Various methods have been developed for their production, encompassing techniques such as solution replacement [5], electrochemical electrolysis [6], ultrasonication [7], and hydrodynamic cavitation [8,9]. One of the main reasons for the preparation of nanobubbles using ultrasound is that sound waves cause liquid to fluctuate between negative and positive pressure, where the gas nuclei in the water form and grow during the negative pressure period and then contract during the favorable pressure period. The bubbles burst if they exceed a critical size [10]. Ultrasonic bubble formation occurs away from the pressure wave belly. The resulting bubbles form in filamentous dendritic clusters, which are, in turn, surrounded by many tiny bubbles that repeatedly clump together and break up [11]. Under the action of cavitation and primary Bjerknes forces in the ultrasonic field, the gas nuclei on the particle surface gradually grow, aggregate, and then form tiny bubbles, driving the particles to aggregate at the acoustic pressure nodes [12]. In a study by Chen et al. [13], it was observed that, under 200 kHz high-intensity focused ultrasound (HIFU), only stable cavitation bubbles were generated on the hydrophobic silica surface. This in turn contributed to particle agglomeration, while no accumulation was formed in the hydrophilic silica suspension.

With the in-depth study of nanobubbles, many research results have shown [14–17] that nanobubbles with large specific surface area and high surface energy are more selective than general bubbles and easily attach to hydrophobic solid surfaces [13,18]. When nanobubbles approach proximity, they establish nanobubble bridges [19], a phenomenon that fosters the agglomeration of microfine mineral particles, bolstering the flotation uplift process. Chen et al. [20] harnessed ultrasonic standing waves operating at 200 kHz to treat finely grained coal, inducing cavitation bubbles to form on the surfaces of hydrophobic coal particles. If the size of the cavitation bubble is larger than the resonance radius, bubble-filled coal particles move to the nodes of the ultrasonic standing wave due to an acoustic radiation force, leading to the rapid aggregation of fine coal particles. Micro-flotation results show that flotation efficiency is significantly improved after ultrasonic standing wave treatment.

Mitra et al. [21] undertook an investigation involving a suspension of hydrophobic glass microspheres characterized by larger particle sizes, employing periodic ultrasonic pulses to do so. The ultrasonic pulses caused the creation of many cavitation bubbles in the liquid in the acceptable size range. It was observed that these bubbles formed particle clusters with the glass microspheres, which attached to relatively large carrier bubbles to form stable bubble–particle aggregates that subsequently floated upwards. Fan et al. [22] found that nanobubbles significantly reduced the bubble rise rate, increasing the air inclusion rate and improving coarse-grained phosphate froth flotation. The presence of nanobubbles in the flotation pulp reduced the rinsing velocity of conventional particle-size bubbles, prolonged the bubble–particle sliding time (contact time), and reduced the tangential sliding velocity of particles on the bubble surface, thereby increasing the probability of bubble–particle attachment and decreasing the probability of detachment. Ahmadi et al. [23] delved into an investigation regarding the influence of nanobubbles on the flotation recovery of microfine- and ultrafine-grained chalcopyrite. Their findings demonstrated that the presence of nanobubbles substantially augmented the recovery rates of chalcopyrite fines and ultra-fines by an impressive margin ranging from 16% to 21%. Furthermore, the incorporation of nanobubbles led to remarkable reductions in the dosages of both the trapping agent (by 75%) and the frothing agent (by 50%).

Sobhy and colleagues [24] conducted a study involving the implementation of nanobubble column flotation for coal processing. They improved the flotation recovery of fine-grained coal by 5 to 50% depending on the process operating conditions, and the amount of frother was reduced by 1/3. Tao [25] investigated the use of nanobubbles for the anionic reverse flotation of hematite and found that their effects significantly improved the grade and recovery of the concentrate. Sobhy [26] calculated that the application of nanobubbles increased the kinetic flotation rate constant by 41%, which implied a significant increase in processing capacity. The nanobubbles encapsulated particles and formed aggregates, thus increasing the probability of particle bubble collisions. Furthermore, the induction time of nanobubbles generated on the surface of hydrophobic particles was reduced by 1/2, thereby increasing the attachment probability.

Liao et al. [27] found that micro- and nanobubbles could enhance the flotation of fine-grained monohydrate hard alumina minerals and significantly improve the flotation recovery of fine-grained monohydrate hard alumina. Taghavi et al. [28] found that nanobubbles significantly improved the flotation recovery of pure magnesite. Lei [29] investigated the role of nanobubbles in the coal–kaolin system, finding that nanobubbles can affect the aggregation state and the degree of kaolinite coverage on the surface of coal particles. On the one hand, nanobubbles fostered the formation of kaolin aggregates. Conversely, the presence of a nanobubble layer hindered the coverage of coal particle surfaces by kaolin, leading to the aggregation of kaolin particles on the surface of coal particles with limited coverage.

Etchepare et al. [30] investigated the flotation effect of nanobubbles on hydrophilic $Fe(OH)_3$ colloidal precipitates. Due to the higher lifting force of microbubbles, the flotation rate was faster. Still, the separation effect was inferior to that of nanobubbles, and so nanobubbles were used in combination with microbubbles to solve the problem of slow flotation rate when using nanobubbles alone. They attributed the increased recovery of hydrophilic $Fe(OH)_3$ precipitates to the entrapment of micro/nanobubbles inside the flocs and solid entrainment by water. Zhou et al. [31] used AFM to find that nanobubbles are difficult to nucleate on highly hydrophilic white mica surfaces and can form stably on sufficiently hydrophobic mineral surfaces. Once nanobubbles are stably nucleated on hydrophobic mineral surfaces, the attraction between mineral particles is greatly enhanced, thus promoting the aggregation and flotation of fine-grained minerals. At the same time, due to the high selectivity of nucleation of nanobubbles on solids, high separation efficiency can also be obtained in separating hydrophobic target minerals from hydrophilic politic minerals. Therefore, increasing the temperature during slurry conditioning may be a potential way to improve the flotation performance of minerals. Tsave et al. [32] discovered that the adherence of microbubbles to the surfaces of fine particles facilitated the attachment of standard-sized bubbles, ultimately improving the flotation recovery of these particles. Zhang et al. [33] noted that nanobubbles reduced the amount of trapping agent used, made flotation faster, and led to particles exhibiting significant aggregation behavior and high adhesion probability.

Researchers have made significant strides in leveraging nanobubbles to enhance the recovery of minerals at the microfine particle size [23,24]. However, vein minerals are not negligible in size, while valuable minerals are finely ground. In particular, for quickly muddied magnesium silicate hydrates, many flotation conditions have not reached the delicate grinding process, which still produces many microfine-grain veinlets, seriously affecting the recovery of valuable minerals. Therefore, in this study, the properties of nanobubbles generated via ultrasonication were investigated by preparing nanobubbles and testing the bubbles' stability. And the flotation behavior of serpentine was investigated by combining nanobubbles with large bubbles of the flotation machine. In the flotation of nickel sulfide ore, easily muddied serpentine minerals have a great impact on the flotation process. Therefore, nanobubbles were prepared via the ultrasonic method and their stability was detected to explore the properties of nanobubbles generated by ultrasound. The nanobubbles were then combined with larger bubbles, typically generated by flotation machines, to scrutinize their impact on the flotation behavior of hydrophilic serpentine. The aim was to examine the influence of nanobubbles on the flotation behavior of hydrophilic serpentine and to reveal their role in optimizing the flotation mechanism of microfine-grained refractory ores.

2. Materials and Methods

2.1. Materials

Serpentine, chemical formula $Mg_6[Si_4O_{10}](OH)_8$, is often divided into three types: lizardite, antigorite, and chrysotite. The serpentine ore used in this study was taken from the Xiuyan jade mine in Liaoning Province, China, and the results of X-ray diffraction (XRD) analysis are shown in Figure 1. Analysis of the sample was conducted using a D/Max-IIIA X-ray diffractometer (Malvern Panalytical, Shanghai, China) with Cu Kα radiation at 40 kV and 30 mA and with a scanning rate of 15 (°)/min from 5° to 70°.

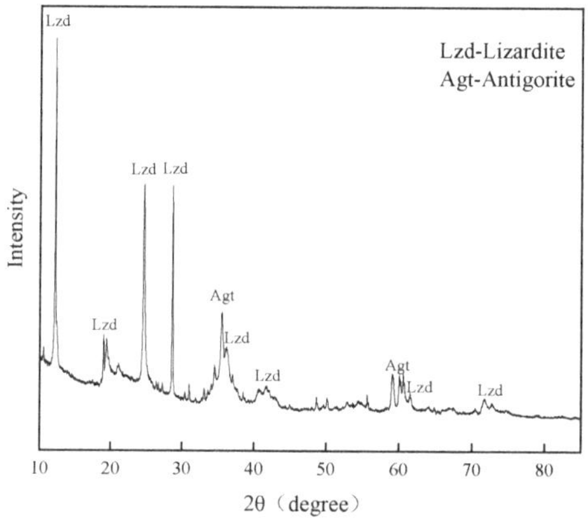

Figure 1. XRD diffraction pattern of serpentine.

Figure 1 shows that the main components of this serpentine ore are lizardite and antigorite minerals. The samples were sent for chemical multi-element analysis. The minerals were uniformly represented and dissolved samples for ICP detection (ICP-OES, Agilent Technologies Inc., Agilent 700, In Santa Clara, CA, USA). The results are shown in Table 1. The data calculation in the table indicates that the purity of serpentine was found to exceed 96%, thus reaching the level required for pure mineral experiments.

Table 1. Serpentine chemical multi-element analysis results.

Elements	Mg	Si	Al	Cl	Fe	S	Ca
Content (%)	33.94	24.88	0.49	0.37	0.32	0.25	0.16

2.2. Methods

2.2.1. Preparation of Nanobubbles

Nanobubbles were generated by utilizing the LC-1500 ultrasonic crusher, manufactured by Shanghai Bingyue Electronic Instrument Co., Ltd. (Shanghai, China) The detection and characterization of nanobubbles were performed using a BeNano 90 nanoparticle size analyzer, a product by Bettersize Instruments (Dandong, China), which is capable of measuring particles within a size range spanning from 2 nm to 36 μm. The measurement error was less than 1%. Measurement was carried out at 22 °C, with 15 mL of the cavitated solution used for detection. The process was repeated three times for each group of specimens to obtain six measurements. Additionally, the viscosity of the pulp following ultrasonication was assessed using a viscometer, specifically the SNB-2 model from Shanghai Jingtian Electronic Instrument Co., Ltd. (Shanghai, China), which features a wide measuring range of 1 to 6,000,000 mPa·s, allowing for comprehensive viscosity analysis.

2.2.2. Nanobubble Flotation Experiments

The agents used were sodium hydroxide (purchased from Shanghai Macklin Biochemical Technology Co., Ltd., Shanghai, China); AR 95%, used as a pH adjuster; Sodium O-butyldithiocarbonate (obtained from Beijing Chemical Industry Group Co., Ltd., Beijing, China); AR 95%, used as a collector; pine alcohol oil (also sourced from Beijing Chemical Industry Group Co., Ltd., Beijing, China); AR 95%, used as a frother; and sodium carboxymethyl cellulose (CMC), used as a depressant (supplied by Shanghai Macklin Biochemical Technology Co., Ltd., Shanghai, China). The ore samples used in the flotation experiments were subjected to crushing, grinding, and sieving processes, yielding five distinct particle-size fractions: 45–74 μm, 38–45 μm, 30–38 μm, 15–30 μm, and <15 μm.

Since a lot of research has been conducted using ultrasonic pretreatment slurry flotation, we know that the main effects are the cavitation effect and the acoustic radiation force effect. The cavitation effect causes the precipitation of dissolved gases in the liquid to form bubbles. Conversely, the acoustic radiation force effect is observed in flotation through actions such as mineral sludge package removal, oxide film detachment, and agent dispersion [20,21]. The objective of this experiment was to investigate the interactions between micro- and nanobubbles and mineral flotation. For this purpose, the experimental procedure involved subjecting 20 mL of a 100 mL pure water sample to ultrasonic treatment via the LC-1500 ultrasonic crusher, employing fixed-duration ultrasound, as depicted in Figure 2.

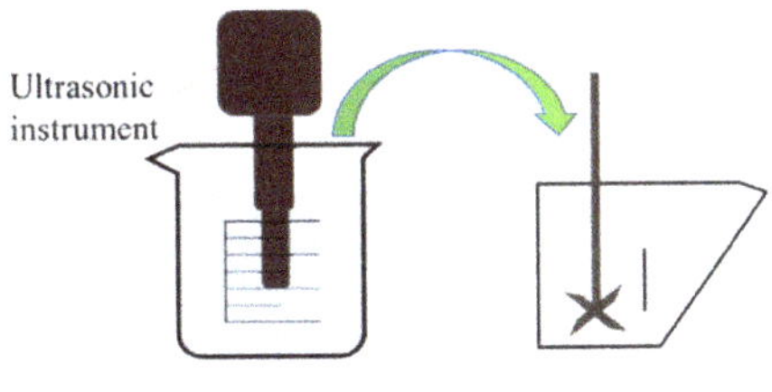

Figure 2. Schematic diagram of micro- and nanobubble flotation.

The flotation experiments were carried out in an XFG-type hanging tank flotation machine (manufactured by Jilin Exploration Machinery Plant, Changchun, China, XFGII5). We took 2.0 g of sample each time, and added 20 mL of pure water (sourced from PINE-TREE XYF2-40-H, Beijing Xianshunyuan Technology Co., Ltd., Beijing, China) after sonication; we then adjusted the pH to 10 using NaOH, and stirred for 3 min. Next, 15 mg/L of sodium O-butyldithiocarbonate and 1×10^{-4} mol/L of pine alcohol oil were added as a collector and frother, respectively, and the flotation was carried out for 3 min. The float and sink products were collected, dried, and weighed, and the flotation recovery was calculated. The flotation machine speed was 1600 r/min.

3. Results and Discussion

3.1. Nanobubble Generation via Ultrasonication

To explore the most favorable conditions for the generation of micro- and nanobubbles through sonication, a series of experiments were conducted involving the sonication of pure water for varying durations: 1 min, 2 min, 5 min, and 10 min, with a control group maintained at 0 min of sonication. These experiments were carried out at a temperature of 25 °C. Each experimental group underwent three repetitions, and the outcomes are graphically depicted in Figure 3. Furthermore, the influence of different ultrasonic durations on the stability of the slurry and its viscosity was investigated, with the results presented in Figure 4.

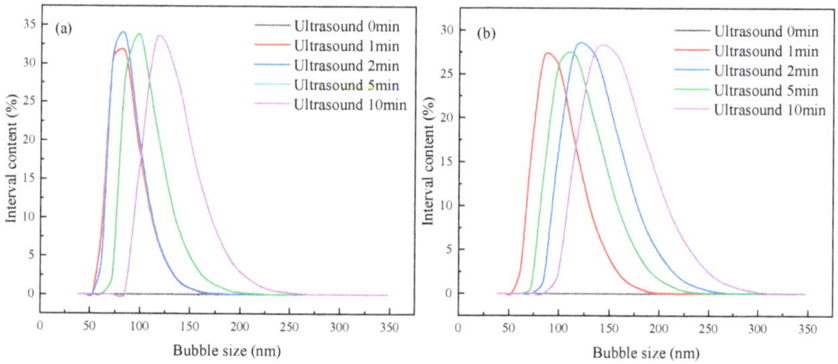

Figure 3. Effect of ultrasonic time on the size distribution of nanobubbles. (**a**) Number distribution; (**b**) volume distribution.

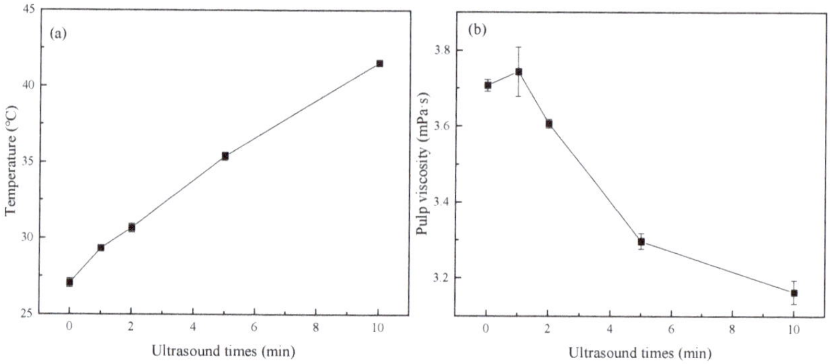

Figure 4. Effect of ultrasonic time on the (**a**) temperature; (**b**) pulp viscosity.

Figure 3a illustrates the presence of numerous nanobubbles within the solution subsequent to sonication, with their diameters falling within the range of approximately 50 to 250 nm. As the sonication duration increased, the size of the bubbles increased, and the size of the nanobubbles produced by 1 min and 2 min of sonication was the smallest. Most of the bubbles were about 80 nm in size. The bubbles produced by 5 min of sonication were about 100 nm, while the bubbles generated following 10 min of sonication were approximately 120 nm in size. Figure 3b provides insights into the volume content distribution of bubbles generated at different sonication durations. Figure 3b also shows that the volume contents of bubbles produced by 1 min, 2 min, 5 min, and 10 min of sonication are similar in size and are all about 27%. In contrast, the bubbles produced via 1 min sonication have the smallest size. The smaller the measurements taken when the volumes are the same, the larger the number of bubbles will be. As such, the bubbles produced via 1 min of sonication have the smallest size and are most numerous [34]. Figure 4a shows that, as ultrasonic time increased, the temperature of the slurry gradually rose too. The results in Figure 4b show that, as ultrasonic time increased, the viscosity of the slurry showed a downward trend. The increasing sonication time raised the water temperature, thereby presenting a higher probability of bubble coalescence and leading to larger sizes and volumes of nanobubbles.

The stability of nanobubbles was investigated, and the nanobubble-containing solutions produced via sonication were left for 0 min, 20 min, 40 min, and 60 min after 1 min of sonication. Then, the bubble sizes were measured, and the results are shown in Figure 5.

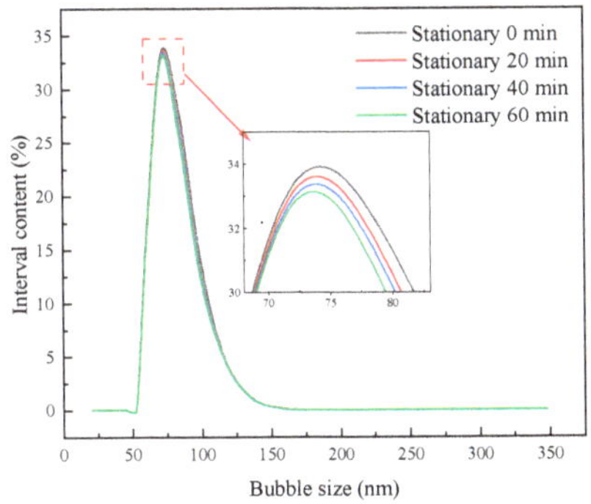

Figure 5. Stability curves of bubbles produced by ultrasonication for 1 min.

The bubble stability curve clearly illustrates that the size of nanobubble generated through ultrasound did not change much with the increasing time. The overall content was stable, which indicated that the nanobubbles were more durable. The produced nanobubbles did not rupture as quickly as conventional bubbles, and some researchers found that most of the nanobubbles present in the solution still existed stably after three months [35].

3.2. Nanobubble Flotation Experiments

The nanobubble flotation experiments were performed by introducing the nanobubble solution generated using ultrasound into the flotation machine. To investigate the impact of the flotation machine on nanobubbles produced through ultrasonic treatment, the distribution of bubbles in water samples was measured 1 min after ultrasonication, at the same time, the gas distribution of the ultrasonic water sample after 9 min of stirring by the flotation machine was measured. It can be seen from Figure 6 that the nanobubbles generated following ultrasonication fell within the size range of approximately 50 150 nm. Additionally, subsequent to agitation in the flotation machine, their size increased to approximately 75–200 nm. Mechanical agitation also caused the aggregation of small-sized bubbles. The bubbles are still within the nanometer size, so the effect of the flotation machine on the flotation of nanobubbles is small. In mechanical flotation, strong agitation could also generate nanobubbles by hydrodynamic cavitation. Therefore, nanobubbles generated via ultrasound and hydrodynamic cavitation co-exist in the system.

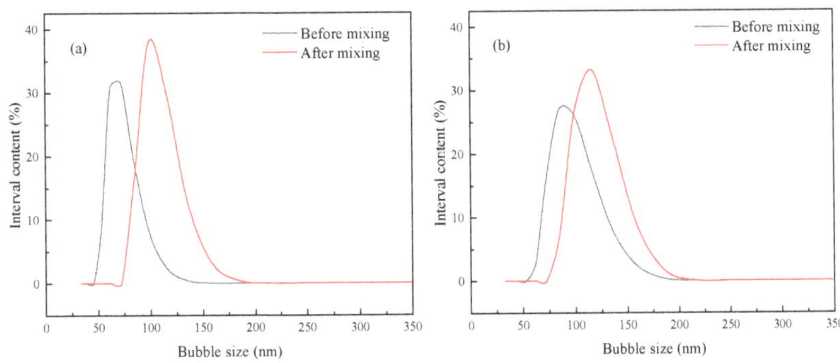

Figure 6. Effect of flotation machine agitation on nanobubbles. (**a**) Number distribution; (**b**) Volume distribution.

In order to explore the influence of nanobubbles on serpentine's flotation behavior, the serpentine recovery rate after ultrasonic pretreatment of the aqueous solution was studied using nanobubbles generated via ultrasonic cavitation combined with large bubbles generated using the hanging tank flotation machine. The flotation conditions were as follows: pH = 9.5, sodium O-butyldithiocarbonate 15 mg/L, pine alcohol oil 1×10^{-4} mol/L. In the experimental procedure, each set of ultrasonicated solution was allowed to equilibrate to the same temperature it was at before ultrasonication prior to proceeding with the flotation test This temperature equilibration ensured consistency in the experimental conditions.

As can be seen from Figure 7, the flotation recovery of serpentine increases as its particle size decreases, reaching 25% for −15 μm. This trend can be attributed to the increase in the apparent viscosity of the pulp as mineral size decreases, leading to a higher entrainment rate [36]. Although serpentine is a hydrophilic mineral, its slow flotation kinetics, non-selective mucus coverage, and entrainment are carried into the froth layer because the particle size is too small. Ultrasonic treatment can significantly impair the recovery of serpentine, especially for the fine-grained grade. The recovery of serpentine increased slowly with the rise in sonication time. The recovery of serpentine in flotation was lowest after 1 min of sonication because of the small size and large number of bubbles produced during 1 min of sonication. The particle size of microfine-grained serpentine is much larger than that of nanobubbles, and as a hydrophilic mineral it cannot enter nanobubble aggregates, which reduces the entrapment of fine-grained serpentine and thus its recovery. Ultrasound raised the temperature and, in turn, reduced the viscosity of the slurry, thereby reducing the entrainment of fine particles to a certain extent. As Figure 7 shows, the floating amount of serpentine was greater with 10 min ultrasonic was greater than that of the ultrasonic 1 min test group; likewise, the viscosity of slurry was significantly lower with 10 min of ultrasound than with 1 min. These findings indicate that the nanobubbles generated by ultrasound played a major role in reducing the entrainment effect.

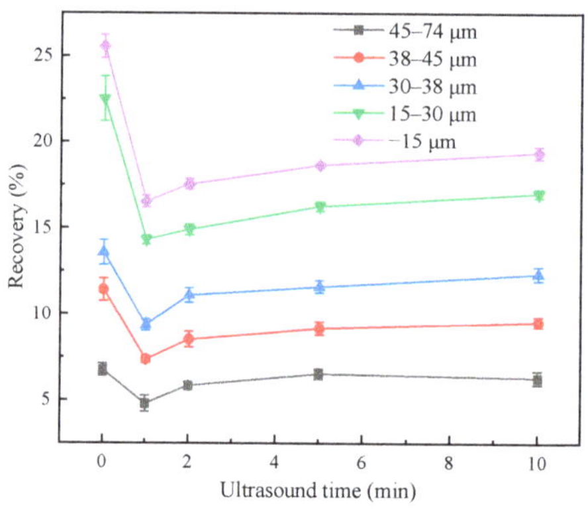

Figure 7. Effect of ultrasonic time on serpentine recovery.

Examination of nanobubble distribution in the concentrate and tailing supernatant after flotation revealed some significant findings, as depicted in Figure 8. In the concentrate, the bubble size ranged from 400 to 1200 nm, while in the tailing it fell within the range of 200 to 500 nm. Notably, these results showed a substantial increase in overall bubble size compared to Figure 5, the results for which were observed after the agitation within the flotation machine. This increase in bubble size suggests that introducing minerals into the system exacerbated the aggregation effect of smaller-sized nanobubbles. It is worth highlighting that the bubble size was substantially larger in the concentrate, and that larger nanobubbles possess enhanced lifting capabilities, making them more likely to rise and become part of the concentrate [30].

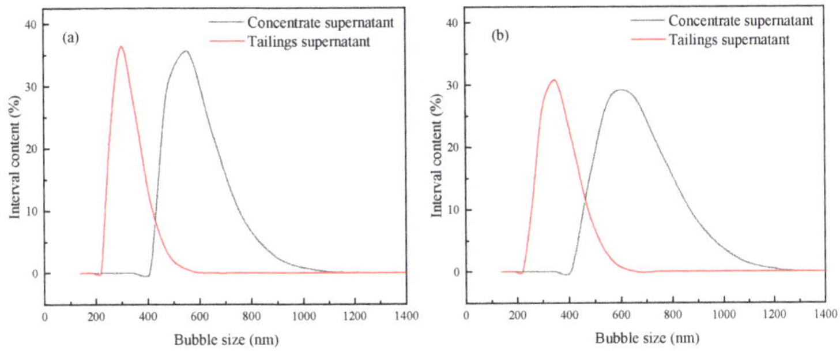

Figure 8. Distribution of nanobubbles in flotation concentrate and tailing. (**a**) Number distribution; (**b**) volume distribution.

Next, nanobubbles were introduced, while depressant CMC was added to the system. Figure 9 illustrates that the presence of nanobubbles lowered the recovery of serpentine across all five particle sizes, with the most significant impact observed for 15–30 μm and −15 μm serpentine fractions. A group with nanobubbles present had a lower recovery with the same amount of added depressant. It has previously been reported that the presence of nanobubbles reduces the amount of depressant required [33]. Consistent with that earlier finding, the recovery level of <15 μm serpentine at a depressant concentration of 40 mg/L was equivalent to the recovery of serpentine at a depressant concentration of 25 mg/L in the presence of nanobubbles. This suggests that nanobubbles can effectively reduce the amount of depressant required to achieve the desired level of serpentine recovery, particularly for fine-grained serpentine fractions.

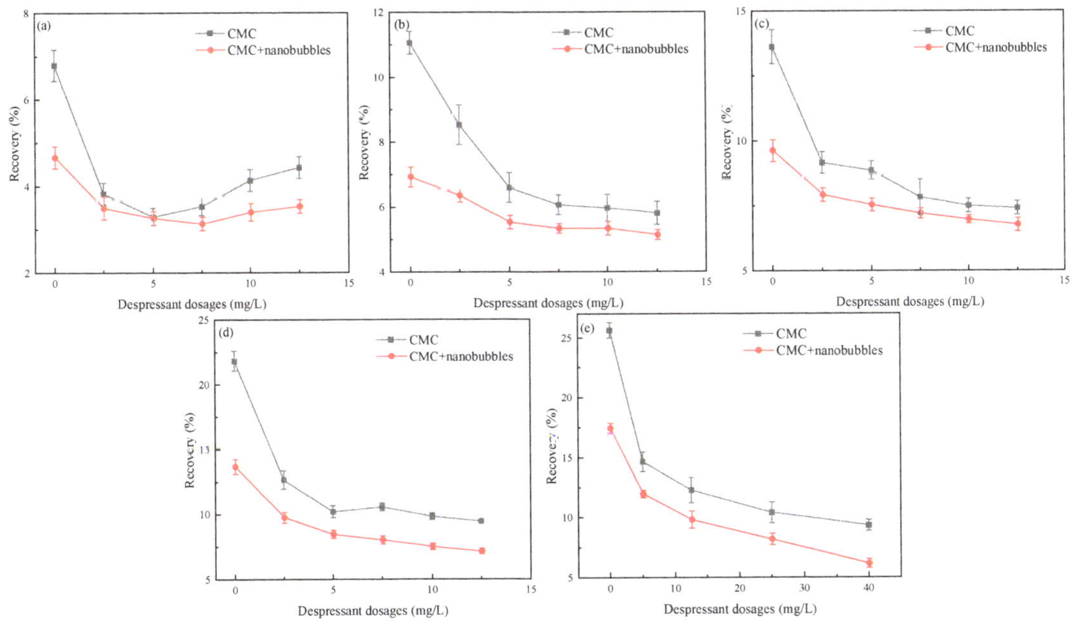

Figure 9. Effect of CMC dosage on the recovery of serpentine in the presence of nanobubbles (pH = 9.5, Sodium O-butyldithiocarbonate 15 mg/L, pine alcohol oil dosage 1×10^{-4} mol/L). (**a**) 45–74 μm; (**b**) 38–45 μm; (**c**) 30–38 μm; (**d**) 15–30 μm; (**e**) −15 μm.

3.3. Analysis of the Mechanism of Nanobubbles to Reduce the Recovery Rate of Microfine-Grained Serpentine

The experimental findings clearly indicate that the presence of nanobubbles reduces the flotation recovery of microfine-grained serpentine. This outcome can be explained by two primary factors. First, the primary mechanisms governing flotation of hydrophilic microfine-grained minerals involve foam entrainment and attachment to the surface of valuable minerals [37]. Given that nanobubbles are significantly smaller than mineral particles, they hinder the adhesion of hydrophilic minerals to valuable mineral surfaces when coming into contact with them [29]. In essence, the presence of nanobubbles disrupts the attachment process, reducing the recovery of hydrophilic serpentine. Second, while the ultrasound-induced reduction in slurry viscosity does contribute to lowering the floatability of hydrophilic serpentine, the dominant factor influencing serpentine flotation is the presence of nanobubbles generated using ultrasound. These nanobubbles play a central role in limiting the attachment and adhesion of serpentine to valuable mineral surfaces, thereby reducing its overall flotation recovery.

Entrainment can be conceptualized as a two-step process involving the movement of particles from the upper pulp to the bubble and then from the bubble to the concentrate. Entrainment is inevitable for both hydrophobic and hydrophilic minerals and is one of the critical factors affecting the concentrate grade [37]. It is important to note that nanobubbles possess relatively weak flotation capabilities; when nanobubbles alone are present, they do not achieve effective mineral recovery [21,30]. In contrast, conventional-sized bubbles exert more significant lifting forces and facilitate faster flotation, as illustrated in Figure 10. During the flotation process, a portion of nanobubbles attaches to the surface of conventional-size bubbles to float, and the domain will exist in the water layer between large bubbles. In addition, the actual volume of nanobubbles produced under natural conditions is very small (gas/bubble holdup <<1% of water), and therefore their effect on water entrainment is negligible. When a large number of nanobubbles exist in the water layer between two large-sized bubbles, that layer has a stronger hydrophobic force on fine-grained hydrophilic minerals; the resulting exclusion of these minerals from the water layer also effectively reduces the entrainment effect.

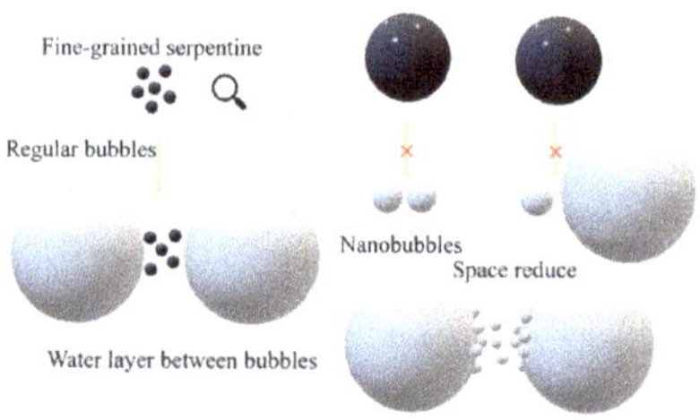

Figure 10. Schematic diagram of nanobubbles reducing the entrapment effect of microfine-grained serpentine.

4. Conclusions

This study investigated the flotation behavior of nanobubbles on the microfine-grained, easily slimed mineral serpentine and yielded the following conclusions:

(1) Nanobubbles were successfully generated through sonication, and we found that the bubble sizes produced via sonication for 1 min and 2 min were the smallest. The stability of the nanobubbles produced by ultrasonication was also studied, and it was found that the nanobubbles were stable. Their size remained consistent, and only a slight decrease in their number was observed with increasing resting time.

(2) Single-mineral flotation experiments were conducted on serpentine using nanobubbles generated through sonication, and we found that the presence of nanobubbles significantly reduced the recovery of serpentine, and that the lowest recovery of

serpentine was obtained by sonication for 1 min. The reduction in serpentine flotation recovery in the presence of nanobubbles was mainly due to a decrease in the froth entrainment rate of serpentine at the microfine-grain level.

(3) Nanobubbles also reduced the amount of depressant required. With the addition of depressant under conditions of 1 min sonication, the recovery level of serpentine in the group without sonication but with 40 mg/L of depressant was equivalent to that of serpentine in the sonicated group, which required only 25 mg/L of depressant to achieve a similar recovery level.

These findings underscore the potential of nanobubbles in mineral flotation processes, particularly for microfine-grained minerals like serpentine, where they can reduce recovery, depressant usage, and froth entrainment.

Author Contributions: Conceptualization, W.X., W.L. and B.L.; methodology, W.X., W.L. and C.L.; validation, W.X., W.L., J.Z. and K.L.; formal analysis, W.X. and Y.S.; investigation, W.X. and C.L.; writing—original draft preparation, W.X. and B.L.; writing—review and editing, W.L. and Y.S.; visualization, B.L. and W.X.; supervision, W.L. and X.S.; project administration, W.L., B.L. and C.L. All authors have read and agreed to the published version of the manuscript.

Funding: This work was financially supported by the National Key Research and Development Project (No. 2022YFC2904202), and the Enterprise entrusted research on Microbubble Flotation Technology of Low-Grade Ore.

Data Availability Statement: Not applicable.

Conflicts of Interest: The authors declare no conflict of interest.

References

1. Feng, D.; Aldrich, C. Effect of particle size on flotation performance of complex sulphide ores. *Miner. Eng.* **1999**, *12*, 721–731. [CrossRef]
2. Ma, Y.; Tao, D.; Tao, Y.; Liu, S. An innovative flake graphite upgrading process based on HPGR, stirred grinding mill, and nanobubble column flotation. *Int. J. Min. Sci. Technol.* **2021**, *31*, 1063–1074. [CrossRef]
3. Zhou, W.; Ou, L.; Shi, Q.; Feng, Q.; Chen, H. Different flotation performance of ultrafine scheelite under two hydrodynamic cavitation modes. *Minerals* **2018**, *8*, 264. [CrossRef]
4. Tao, D. Recent advances in fundamentals and applications of nanobubble enhanced froth flotation: A review. *Miner. Eng.* **2022**, *183*, 107554. [CrossRef]
5. Liu, M.; Zhao, W.; Wang, S.; Guo, W.; Tang, Y.; Dong, Y. Study on nanobubble generation: Saline solution/water exchange method. *ChemPhysChem* **2013**, *14*, 2589–2593. [CrossRef]
6. Guan, M.; Guo, W.; Tang, Y.; Hu, J.; Dong, Y. Investigation on the temperature difference method for producing nanobubbles and their physical properties. *ChemPhysChem* **2012**, *13*, 2115–2118. [CrossRef] [PubMed]
7. Cho, S.H.; Kim, J.Y.; Chun, J.H.; Kim, J.D. Ultrasonic formation of nanobubbles and their zeta-potentials in aqueous electrolyte and surfactant solutions. *Colloids Surf. A Physicochem. Eng. Asp.* **2005**, *269*, 28–34. [CrossRef]
8. Nirmalkar, N.; Pacek, A.W.; Barigou, M. On the existence and stability of bulk nanobubbles. *Langmuir* **2018**, *34*, 10964–10973. [CrossRef]
9. Etchepare, R.; Oliveira, H.; Nicknig, M.; Azevedo, A.; Rubio, J. Nanobubbles: Generation using a multiphase pump, properties and features in flotation. *Miner. Eng.* **2017**, *112*, 19–26. [CrossRef]
10. Suslick, K.S.; Doktycz, S.J.; Flint, E.B. On the origin of sonoluminescence and sonochemistry. *Ultrasonics* **1990**, *28*, 280–290. [CrossRef] [PubMed]
11. Hatanaka, S.; Yasui, K.; Kozuka, T.; Tuziuti, T.; Mitome, H. Influence of bubble clustering on multibubble sonoluminescence. *Ultrasonics* **2002**, *40*, 655–660. [CrossRef] [PubMed]
12. Chen, Y.; Ni, C.; Xie, G.; Liu, Q. Toward efficient interactions of bubbles and coal particles induced by stable cavitation bubbles under 600 kHz ultrasonic standing waves. *Ultrason. Sonochem.* **2020**, *64*, 105003. [CrossRef]
13. Chen, Y.; Zheng, H.; Truong, V.N.T.; Xie, G.; Liu, Q. Selective aggregation by ultrasonic standing waves through gas nuclei on the particle surface. *Ultrason. Sonochem.* **2020**, *63*, 104924. [CrossRef] [PubMed]
14. Nazari, S.; Shafaei, S.Z.; Gharabaghi, M.; Ahmadi, R.; Shahbazi, B.; Fan, M. Effects of nanobubble and hydrodynamic parameters on coarse quartz flotation. *Int. J. Min. Sci. Technol.* **2019**, *29*, 289–295. [CrossRef]
15. Ding, S.; Xing, Y.; Zheng, X.; Zhang, Y.; Cao, Y.; Gui, X. New insights into the role of surface nanobubbles in bubble-particle detachment. *Langmuir* **2020**, *36*, 4339–4346. [CrossRef] [PubMed]
16. Cui, R. Study on the Improvement of Low-Rank Coal and Its Semi-Coke Adsorption on Congo Red Dye by Micro-Nano Bubbles. Master's Thesis, China University of Mining and Technology, Xuzhou, China, 2020.

17. Ma, F.; Zhang, P.; Tao, D. Surface nanobubble characterization and its enhancement mechanisms for fine-particle flotation: A review. *Int. J. Miner. Metall. Mater.* **2022**, *29*, 727–738. [CrossRef]
18. Azevedo, A.; Etchepare, R.; Calgaroto, S.; Rubio, J. Aqueous dispersions of nanobubbles: Generation, properties and features. *Miner. Engr.* **2016**, *94*, 29–37. [CrossRef]
19. Hampton, M.A.; Nguyen, A.V. Nanobubbles and the nano-bubble bridging capillary force. *Adv. Colloid Interface Sci.* **2010**, *154*, 30–55. [CrossRef]
20. Chen, Y.; Xie, G.; Chang, J.; Grundy, J.; Liu, Q. A study of coal aggregation by standing-wave ultrasound. *Fuel* **2019**, *248*, 38–46. [CrossRef]
21. Mitra, S.; Hoque, M.M.; Evans, G.; Nguyen, A.V. Direct visualisation of bubble-particle interactions in presence of cavitation bubbles in an ultrasonic flotation cell. *Miner. Eng.* **2021**, *174*, 107258. [CrossRef]
22. Fan, M.; Tao, D.; Rick, H.; Luo, Z. Nanobubble generation and its applications in froth flotation (part II): Fundamental study and theoretical analysis. *Min. Sci. Technol. Chin.* **2010**, *20*, 159–177. [CrossRef]
23. Ahmadi, R.; Khodadadi, D.A.; Abdollahy, M.; Fan, M. Nano-microbubble flotation of fine and ultrafine chalcopyrite particles. *Int. J. Min. Sci. Technol.* **2014**, *24*, 559–566. [CrossRef]
24. Sobhy, A.; Tao, D. Nanobubble column flotation of fine coal particles and associated fundamentals. *Int. J. Miner. Process* **2013**, *124*, 109–116. [CrossRef]
25. Tao, D.; Wu, Z.; Sobhy, A. Investigation of nanobubble enhanced reverse anionic flotation of hematite and associated mechanisms. *Powder Technol.* **2021**, *379*, 12–25. [CrossRef]
26. Sobhy, A. Cavitation Nanobubble Enhanced Flotation Process for More Efficient Coal Recovery. Master's Thesis, University of Kentucky, Lexington, KY, USA, 2013.
27. Liao, S.; Ou, M.; Zhou, W. Micro-nano bubbles properties induced by hydrodynamic cavitation and their influences on fine mineral flotation. *Chin. J. Nonferrous. Met.* **2019**, *29*, 1567–1574.
28. Tsave, P.K.; Kostoglou, M.; Karapantsios, T.D.; Lazaridis, N.K. A Hybrid Device for Enhancing Flotation of Fine Particles by Combining Micro-Bubbles with Conventional Bubbles. *Minerals* **2021**, *11*, 561. [CrossRef]
29. Lei, W. Study on the Preparation of Nano-Bubbles and Their Effect on Coal Slime Flotation. Master's Thesis, Wuhan University of Science and Technology, Wuhan, China, 2020.
30. Etchepare, R.; Azevedo, A.; Calgaroto, S.; Rubio, J. Removal of ferric hydroxide by flotation with micro and nanobubbles. *Sep. Purif. Technol.* **2017**, *184*, 347–353. [CrossRef]
31. Zhou, W.; Wu, C.; Lv, H.; Zhao, B.; Liu, K.; Ou, L. Nanobubbles heterogeneous nucleation induced by temperature rise and its influence on minerals flotation. *Appl. Surf. Sci.* **2020**, *508*, 145282. [CrossRef]
32. Taghavi, F.; Noaparast, M.; Pourkarimi, Z.; Nakhaei, F. Comparison of mechanical and column flotation performances on re-covery of phosphate slimes in presence of nano-microbubbles. *J. Cent. South Univ.* **2022**, *29*, 102–115. [CrossRef]
33. Zhang, Z.; Ren, L.; Zhang, Y. Role of nanobubbles in the flotation of fine rutile particles. *Miner. Eng.* **2021**, *172*, 107140. [CrossRef]
34. Mo, C. Research on the Generation Method and Properties of Nanobubble Based on Ultrasonic Cavitation. Master's Thesis, University of Chinese Academy of Sciences (Shanghai Institude of Applied Physics, Chinese Academy of Science), Shanghai, China, 2019.
35. Nirmalkar, N.; Pacek, A.W.; Barigou, M. Bulk nanobubbles from acoustically cavitated aqueous organic solvent mixtures. *Langmuir* **2019**, *35*, 2188–2195. [CrossRef] [PubMed]
36. Zou, Y. Fundamental Study on the Effect of Pulp Viscosity on Gangue Entrainment in Chalcopyrite Flotation. Master's Thesis, China University of Mining and Technology, Xuzhou, China, 2021.
37. Wang, C.; Sun, C.; Liu, Q. Entrainment of Gangue Minerals in Froth Flotation: Mechanisms, Models, Controlling Factors, and Abatement Techniques—A Review. *Min. Reclam. Environ.* **2021**, *38*, 673–692. [CrossRef]

minerals

MDPI

Article

The Prospect of Recovering Vanadium, Nickel, and Molybdenum from Stone Coal by Using Combined Beneficiation and Metallurgy Technology Based on Mineralogy Features

Miao Wang [1,2,3,4,5], Liulu Cai [1,3,5], Jiankang Wen [1,3,4,5,*], Wenjuan Li [1,3,4,5], Xinlong Yang [1,3,5] and Hongying Yang [2]

1 National Engineering Research Center for Environment-Friendly Metallurgy in Producing Premium Non-Ferrous Metals, GRINM Group Corporation Limited, Beijing 101407, China
2 School of Metallurgy, Northeastern University, Shenyang 110891, China
3 GRINM Resources and Environment Tech. Co., Ltd., Beijing 101407, China
4 General Research Institute for Nonferrous Metals, Beijing 100088, China
5 GRIMAT Engineering Institute Co., Ltd., Beijing 101407, China
* Correspondence: kang3412@126.com

Abstract: Black shale ore contains rich strategic metal resources such as vanadium, nickel, and molybdenum, but due to its complex composition, it is currently only used in the vanadium extraction industry. Metals such as nickel and molybdenum have not been effectively recovered, resulting in environmental pollution and resource waste. Using mineralogical features and a combination of beneficiation and metallurgy-based tests, the present work carried out feasibility studies of the combined beneficiation and metallurgy processes. The mineralogical features of the stone coal sample were studied using chemical analysis, an automatic mineral analyzer (BPMA), etc., and we identified the main phase composition, embedded characteristics, and particle size distribution of the associated strategic metals, vanadium, nickel, and molybdenum. The results showed that the grade of V_2O_5 in the stone coal was 1.29%, which was mainly present in carbonaceous clay and mica minerals. The nickel grade was 0.53%, mainly in the form of nickel–magnesium spinel and a small amount of nickel-containing magnesite. The stone coal contained 0.11% molybdenum; the mineral particles were fine, mostly in the form of molybdenite, and some were associated with carbonaceous matter and carbonaceous clay minerals. Based on the mineralogical feature, we proposed using the scrubbing–desliming and flotation process to enrich vanadium, nickel, and molybdenum. Our preliminary experiments obtained two products: vanadium–molybdenum-rich sludge and nickel-containing tailings. The V_2O_5 and molybdenum grades in the sludge were 4.10% and 0.44%, respectively, and the recovery was 41.31% and 51.40%, respectively. The nickel grade in the tailings was 1.49%. These products were roasted and leached. The vanadium, nickel, and molybdenum in the stone coal were effectively recovered through the beneficiation–metallurgy combination process, and the comprehensive utilization rate of the stone coal was improved.

Keywords: stone coal; mineralogy; beneficiation–metallurgy combination process; vanadium-nickel-molybdenum recovering

Citation: Wang, M.; Cai, L.; Wen, J.; Li, W.; Yang, X.; Yang, H. The Prospect of Recovering Vanadium, Nickel, and Molybdenum from Stone Coal by Using Combined Beneficiation and Metallurgy Technology Based on Mineralogy Features. *Minerals* **2023**, *13*, 21. https://doi.org/10.3390/min13010021

Academic Editor: Hyunjung Kim

Received: 1 November 2022
Revised: 15 November 2022
Accepted: 21 December 2022
Published: 23 December 2022

1. Introduction

Vanadium (V), nickel (Ni), and molybdenum (Mo) are all important strategic metal resources and are widely used in steel, electronics, aerospace, the military, industry, and so on [1–3]. Key metals in black shale have received growing attention as global ore grades have continued to decline [4,5]. Stone coal is a vanadium-bearing shale (total organic carbon content > 10%) and a polymetallic symbiotic ore. More than 60 associated elements have been found in stone coal and, usually, the V content is the highest, followed by Mo, Ni, U, P, etc. [6]. China has a total of 61.88 billion tons of stone coal reserves, out of which there are 118 million tons of V_2O_5 reserves, which is the main raw materials used for vanadium production, apart from vanadium titanomagnetite [7]. China started researching methods of extracting vanadium from stone coal in the 1970s, and over time, a series of achievements were made regarding the research and development of metallurgical vanadium extraction processes and mechanisms [8,9]. However, the contents of the associated nickel and molybdenum found in stone coal are generally low and difficult to recover. At present, there are few studies on the recovery of nickel and molybdenum from stone coal.

The occurrence states of vanadium, nickel, and molybdenum in stone coal in different regions are complex and diverse, and the combined beneficiation and smelting processes used to extract them also vary. According to the literature, vanadium in stone coal mainly exists in the form of V^{3+} in minerals such as muscovite ($KAl_2[AlSi_3AlO_{10}](OH,F)_2$) and kaolinite ($Al_2Si_2O_5(OH)_4$), and it replaces Al^{3+}, Fe^{3+}, etc., which exist in the four 4-coordinate silicon–oxygen tetrahedron "multiple network layer" and the six-coordinate aluminum–oxygen octahedron "single network layer" [10,11]. A small amount of V, including V(IV) and V(V), adsorbs on the surface of organic matter, iron oxide, or clay minerals, or exists in the form of independent minerals such as calcium vanadium garnet ($Ca_3V_2[SiO_4]_3$) and vanadium uranite [10,12]. Nickel and molybdenum in stone coal mostly exist as sulfides. Mineralogical feature studies have shown that nickel-bearing minerals include vaesite (Ni_2S), polydymite (Ni_3S_4), gersdorffite (NiAsS), millerite (NiS), and pyrite containing nickel ($(FeNi)_9S_8$) (colloidal pyrite). Molybdenum-bearing minerals include molybdenite and carbonaceous-sulfide–sulfur–molybdenum aggregates [13,14]. Due to the different occurrence states of vanadium and the mineral composition characteristics of stone coal in various regions, there have been many studies on the combined beneficiation–metallurgy process. For high-carbon stone coal, Wang et al. [15] adopted a combined flotation–decarburization–flotation vanadium-leaching process. The V_2O_5 grade was increased from 0.92% to 1.32% with a recovery of 88.38%, and the leaching efficiency of the vanadium concentrate was 85%. Carbon concentrate obtained with flotation can be used as fuel, and the leaching amount and production costs can be reduced by using this process. For calcareous stone coal, Liu [16] adopted a combined roasting–decarburization–desliming–flotation–leaching process to obtain V_2O_5 concentrate with a grade of 1.07% and a recovery of 83.30%. In total, 62.52% of the CaO was removed, and the leaching efficiency of the vanadium concentrate increased by 13.53% compared with the direct leaching of raw ore. The combined beneficiation and metallurgy process can effectively reduce the ore-processing volume and acid consumption of vanadium extraction from stone coal. For siliceous stone coal, Gu et al. [17] enriched vanadium-bearing mica and clay-mineral-absorbing vanadium using a preferential decarburization–desliming–flotation process. After a locked-circuit process of separation and smelting, the V_2O_5 grade of the mixed vanadium concentrate was 2.04% and the recovery was 83.41%. The process eliminated 64.87% of gangue minerals. In addition to vanadium recovery, some researchers have studied the comprehensive recovery of nickel, molybdenum, and other associated elements from stone coal. Liu et al. [18] proposed a new technology using flotation–roasting–alkali-leaching. Using flotation to extract molybdenum, the molybdenum concentrate grade was increased from 0.036% in the head ore to 0.112%, and the recovery was 81.3%. The molybdenum extraction tailings were treated by the calcium-roasting–alkali leaching process, and the vanadium leaching efficiency reached over 80% with the overall recovery of 78.2%. This process has proceeded to the stage of a small-scale application. At present, stone coal is still mainly used to extract vanadium, and the recovery of nickel and molybdenum is less frequently studied. Research on the occurrence states of nickel and molybdenum in stone coal is not comprehensive, and it is still difficult to effectively guide the separation and extraction of nickel and molybdenum from stone coal. A mineralogical feature study of nickel and molybdenum in stone coal may provide a foundation for the efficient utilization of nickel and molybdenum resources.

Determining the beneficiation–metallurgy process is mainly performed by finding the mineral composition of ore samples and the occurrence states of the target minerals.

The key to the comprehensive utilization of vanadium, nickel, and molybdenum resources in stone coal is to systematically study the occurrence states of these three metals in the stone coal. By analyzing the chemical and mineral composition, physical properties, dissemination characteristics, and grain size distribution of the ore, one can obtain basic information to support the feasibility study of beneficiation and metallurgy processes [19]. Comprehensively recovering vanadium and the strategic metals nickel and molybdenum from stone coal not only reduces metal resources waste, but also brings certain economic benefits, and it reduces the risk of nickel and molybdenum polluting the environment.

In the present work, a systematic mineralogical feature study on stone coal from Shaanxi Province was carried out. The mineral types, occurrence states, and distribution characteristics of vanadium, nickel, and molybdenum were identified. Based on the study, we propose to use the scrubbing–desliming and flotation process to pre-enrich vanadium, nickel, and molybdenum.

2. Materials and Methods

2.1. *Raw Ore and Reagents*

The stone coal used in this study came from Shaanxi Province. The raw ore was crushed to ~2 mm with a jaw crusher and a double-roller crusher and then mixed. Analysis showed that the organic carbon content in the sample was 6.13%, V_2O_5 was 1.29%, Mo was 0.11%, and Ni was 0.53%. In the flotation process, water glass was used as a dispersant; terpineol was used as a foaming agent; kerosene, butyl xanthate, butylammonium dithiophosphate, and 3030C (cationic surfactant) were used as collectors; sulfuric acid was used as a pH regulator; sulfuric acid was used as an acid-leaching agent; calcium fluoride was used as a leaching aid.

2.2. *Mineralogy Study*

A chemical analysis of V, Ni, Mo, and other elements in the raw ore was carried out using acid digestion—ICP-OES (Agliet 725, Agilent Technologies Inc., Santa Clara, CA, USA) [20], and we obtained the distribution of vanadium, nickel, and molybdenum in the raw ore and its fractions. The main components of the raw ore were identified by X-ray fluorescence (XRF, PANalytical Axios, Almelo, Holland) semi-quantitative analysis. X-ray diffraction (XRD) analysis was conducted using a D/Max-IIIA X-ray diffractometer (Malvern Panalytical, Shanghai, China) with Cu Kα radiation at 40 kV and 30 mA with a scanning rate of 15 (°)/min from 5° to 70°.

We used the BGRIMM Process Mineralogy Analyzer (BPMA, BGRIMM Technology Group, Beijing, China) which consists of a TESCAN VEGA scanning electron microscope, a Bruker QUANTAX 200 dual-probe spectrometer, and automatic analysis software for processing mineralogy (BMPA V2.0) [21]. The stone coal inlay sample was prepared by mixing the ore, epoxy resin, and heated carnauba wax in a mold and then consolidating the particle specimen into blocks through slow hardening of the epoxy resin, followed by grinding and polishing. The sample analysis method was: scanning electron microscope high-vacuum mode, backscattered electron (BSE) probe, acceleration voltage (HV) 20 kV, and working distance (WD) 15.4 mm. The basic principle of the BPMA is to use the system to control the scanning electron microscope to obtain a backscattered electron image that can reflect the different characteristics of the mineral chemical components and combine with the X-ray energy spectrometer data to identify the minerals. This is a rapid and quantitative test of mineral composition and content, particle size, dissociation degree, and other mineralogical parameters in stone coal [22]. Analyses of C were conducted with a high-frequency infrared absorption spectrometer (CS-3000G, NCS Testing Technology Co., Ltd., Beijing, China).

2.3. *Recovery Method of Vanadium, Nickel, and Molybdenum*

(1) Beneficiation experiments

During the scrubbing and desliming test, the slurry concentration was 33%, sodium silicate was 2 kg/t, and the stirring speed was 500 rpm. After stirring for 20 min, water was added to adjust the slurry concentration to 20%. After another 5 min, stirring was stopped, and the slurry was allowed to settle for 5 min. The sludge was filtered and dried to obtain vanadium and molybdenum concentrates. The scrubbed tailings were ground to a fineness of 85%—74 μm, then added with kerosene (300 g/t) and terpenic oil (300 g/t) for decarburization. Next, butyl xanthate (60 g/t), butylammonium dithiophosphate (60 g/t), and terpenic oil (150 g/t) were added for molybdenite flotation. Based on the pH conditions of quartz and mica flotation separation, we added sulfuric acid to adjust the pH of the pulp to 3–4. 3030C (350 g/t) was used to float mica to obtain vanadium mica concentrate. In this study, beneficiation experiments were carried out three times. Average values of the results are reported here. The experimental flow chart is shown in Figure 1.

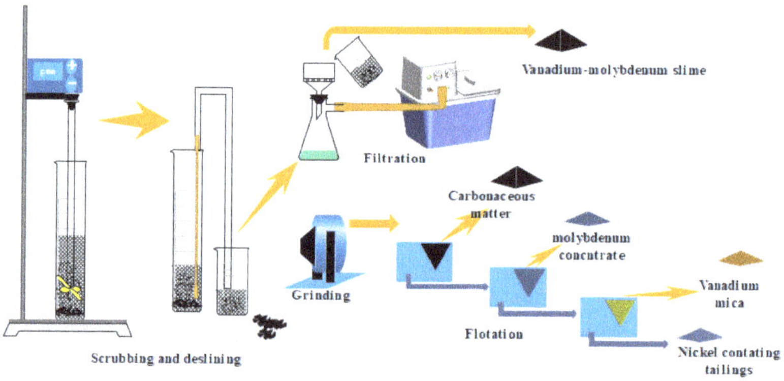

Figure 1. The experimental flow chart of scrubbing–desliming–flotation.

(2) Metallurgy experiments

The vanadium–molybdenum slime, carbonaceous matter, and molybdenum concentrate were roasted and decarburized at 650 °C for 2 h, and then leached with 1.84 mol/L of sulfuric acid for 10 h at a liquid–solid ratio of 4:1. The vanadium-containing mica concentrate and nickel-containing tailings were directly leached under the same condition as the roasted slag above. In this study, leaching experiments were carried out three times, and the average value of the leaching efficiency is presented here.

3. Results and Discussion

3.1. Mineralogy Study

3.1.1. X-ray Fluorescence Semi-Quantitative Analysis and Chemical Analysis

The results are shown in Tables 1 and 2.

Table 1. X-ray fluorescence semi-quantitative analysis results.

Element	Content/%	Element	Content/%	Element	Content/%
SiO_2	56.72	CaO	2.62	ZnO	0.14
SO_3	18.02	Fe_2O_3	2.36	MnO	0.03
V_2O_5	1.16	K_2O	2.07	CuO	0.03
MoO_3	0.09	P_2O_5	0.7	Cl	0.02
NiO	0.52	BaO	0.43	As_2O_3	0.02
Na_2O	5.09	Cr_2O_3	0.41	Co_3O_4	0.01
MgO	5.08	ZrO_2	0.25	PbO	0.01
Al_2O_3	4.05	TiO_2	0.15	SeO_2	0.01

Table 2. Chemical multielement analysis.

Element	Content/%	Element	Content/%	Element	Content/%
V	0.72	K	1.53	Ti	0.05
Mo	0.11	Fe	1.16	Cu	0.018
Ni	0.53	Ca	0.7	Mn	0.014
S	4.89	Si	24.04	Sn	0.013
Mg	4.68	Ba	0.36	As	0.0097
Na	3.41	P	0.28	C	7.02
Al	1.71	Cr	0.20		

According to the results in Tables 1 and 2, the main chemical elements in the sample were Si, S, C, Na, Mg, Al, etc., and the content of SiO_2 in the raw ore was relatively high, followed by Mg and Na. The contents of the valuable elements V, Ni, and Mo were 0.72%, 0.53%, and 0.11%, respectively, all of which reached or exceeded the mining grade, and the comprehensive recovery value was high. The grade of V_2O_5 was 1.29% in the stone coal.

3.1.2. XRD Results

The XRD analysis result is shown in Figure 2. The main minerals of the stone coal were quartz and mica with a small amount of sanidine, nickel oxide, loewite, and pyrite. Because the carbon material was amorphous and the degree of crystallization was poor, the peak value did not appear in the spectrum. Molybdenum mineral did not either due to its low content. V(III) in the stone coal was mostly in the form of lattice substitution in muscovite [10], and no independent V mineral was found in the diffraction analysis.

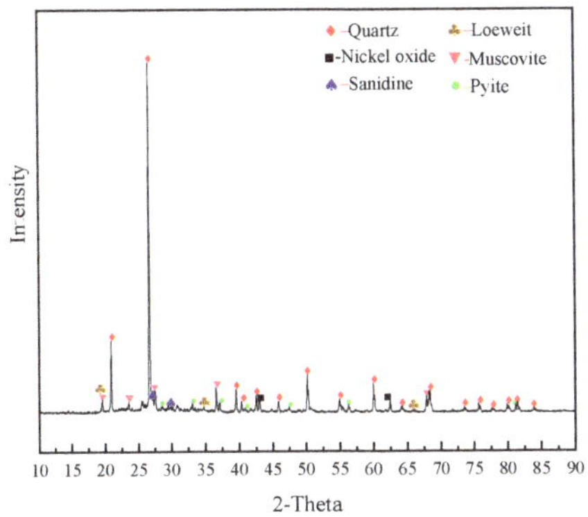

Figure 2. XRD pattern of the stone coal raw ore.

3.1.3. Mineral Composition and Embedded Characteristics

(1) Mineral composition

The main gangue minerals in the sample were quartzes, which made up 38.05% of the content. Sulphate minerals such as loewite and nickel–potassium–magnesium–alum accounted for about 12%. Carbonaceous matter content in the stone coal was 15.87%, some of which existed in the form of organic carbon, and some closely bound to clay minerals. Vanadium-containing minerals included vanadium mica and carbonaceous clay minerals. Nickel-containing minerals were nickel (magnesium, iron, aluminum, and chromium) spinel, and the molybdenum-containing mineral was molybdenite. The mineral composition of the stone coal is shown in Figure 3. The embedding characteristics of the vanadium, nickel, and molybdenum minerals are shown in Figure 4

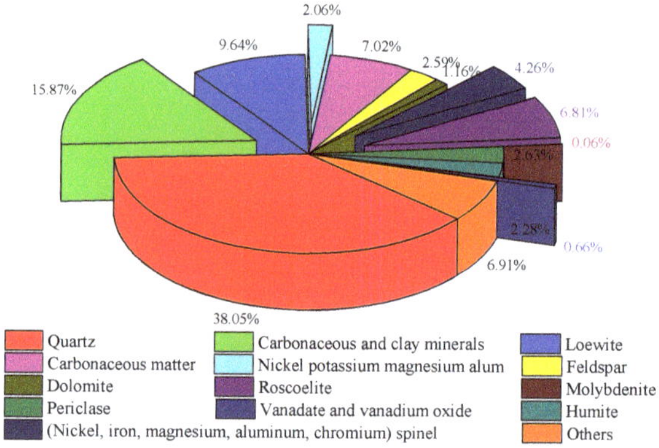

Figure 3. Mineral composition of the stone coal.

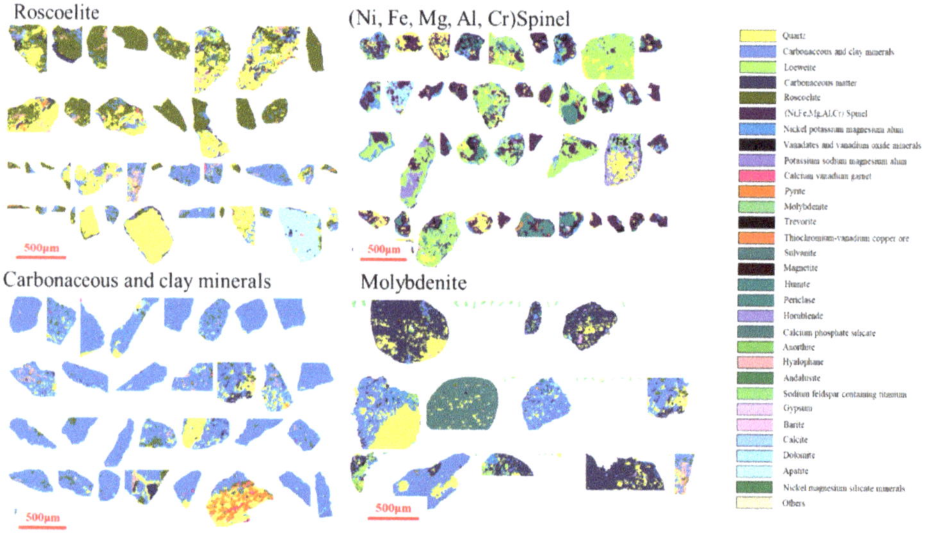

Figure 4. Embedded characteristics of key minerals.

Results of the ore screening are shown in Table 3. The coarse particles yield was high, with +425 μm and +150 μm accounting for 39.37% and 16.39%, respectively. The intermediate particles yield was lower than that of the coarse particles with −150 + 106, −106 + 74, and −74 + 4 μm accounting for less than 10%. The fine particles yield (−45 μm) was 12.51%. From the perspective of metal distribution, vanadium and nickel were mainly distributed in coarse and fine-grained fractions, and the distribution in the intermediate particles was very small. Most of the molybdenum was distributed in micro-fine ores, and the distribution in coarse and intermediate particles was smaller. Based on the metal grades, the nickel grade was low and only in the +425 μm part, and the other grades were similar. The vanadium content in the fine particles was highest, and the grades of the coarse particles and intermediate particles were close. The molybdenum grade increased gradually with decreasing particle size. In general, the distribution rates and grades of the vanadium and molybdenum in the fine-grained ores were high. Vanadium and molybdenum were mainly distributed in fine-grained ores, while nickel was evenly distributed in each grain size. We also found that the raw ore dissolved in the wet screening process. The dissolution process was further analyzed.

Table 3. Screening results for the raw ore.

Fraction/μm	Yield/%	Grade/%			Distribution %		
		V_2O_5	Ni	Mo	V_2O_5	Ni	Mo
+425	39.37	0.97	0.27	0.01	27.23	18.03	3.52
−425~+150	16.39	1.56	0.97	0.01	18.21	27.37	1.22
−150~+106	2.70	1.87	1.63	0.03	3.58	7.56	0.68
−106~+74	3.0	1.14	1.76	0.18	2.41	9.06	4.77
−74~+45	3.44	2.23	1.91	0.59	5.43	11.32	17.69
−45	12.51	4.85	1.24	0.66	43.11	26.66	72.12
Dissolution	22.59	-	-	-	-	-	-
Total	100	1.29	0.53	0.11	100	100	100

(2) Analysis of the vanadium phase and occurrence states

The BPMA's results show that the vanadium minerals in the stone coal included vanadium mica, carbonaceous clay minerals, calcium vanadium garnet, a small amount of vanadium oxide, and sulfur–chromium–vanadium–copper ore. The analysis results of the vanadium phase are shown in Table 4. We can see that the main minerals containing vanadium were vanadium mica and clay minerals, and a small amount was distributed in tourmaline and garnet. Vanadium in clay minerals and mica minerals accounted for 65.64% and 24.43% of the total vanadium, respectively.

Table 4. Phase analysis of vanadium in the stone coal.

Phase	V_2O_5 Content/%	V_2O_5 Distribution %
Micaceous minerals	0.32	24.43
Iron oxides and clay minerals	0.86	65.64
Tourmaline and garnet	0.13	9.92
Total vanadium	1.31	100.00

Subsequently, the dissemination characteristics of the vanadium-containing minerals were studied. Because the optical properties of vanadium-containing minerals in stone coal are similar to those of quartz, feldspar, and other major gangue minerals, it is difficult to distinguish them [23]. The content of vanadium minerals in the samples was small, and it took a long time to search for vanadium minerals using a microscope. Therefore, the automatic analysis system (BPMA) for minerals was used to quickly and intuitively analyze the mineral composition in the stone coal. The distribution of vanadium in the detection area was obtained by using a backscattered electron image combined with an X-ray energy spectrum analysis, and the area with high vanadium content was analyzed. The main vanadium minerals in the stone coal are shown in Figure 5

According to the morphology and dissemination relationship in the vanadium-containing mica in the stone coal listed in Figure 4, it can be seen that the particle size of the vanadium-containing mica was not uniform: there was no regular shape, and most of them were closely associated with quartz, calcium–vanadium garnet, feldspar, carbonaceous materials, and other minerals. Some of the calcium–vanadium garnets were surrounded by vanadium mica. Carbonaceous matter filled the spaces between vanadium mica and quartz in strips. Pyrite was dispersed on the surface of the feldspar and quartz aggregates in the form of micro-fine particles. Most of the vanadium mica and calcium vanadium garnets coexisted, so they could be recovered together using flotation.

In addition to roscoelite, the carbonaceous clay minerals also contained some vanadium. Figure 6 shows the vanadium distribution in the mica and clay minerals. The carbonaceous clay minerals were mostly veined and disseminated among the vanadium-bearing mica, feldspar, and quartz. The X-ray spectrum in Table 5 shows that the mass fraction of V in the carbonaceous clay minerals was about 0.5%, and that in the vanadium mica was 13.33%.

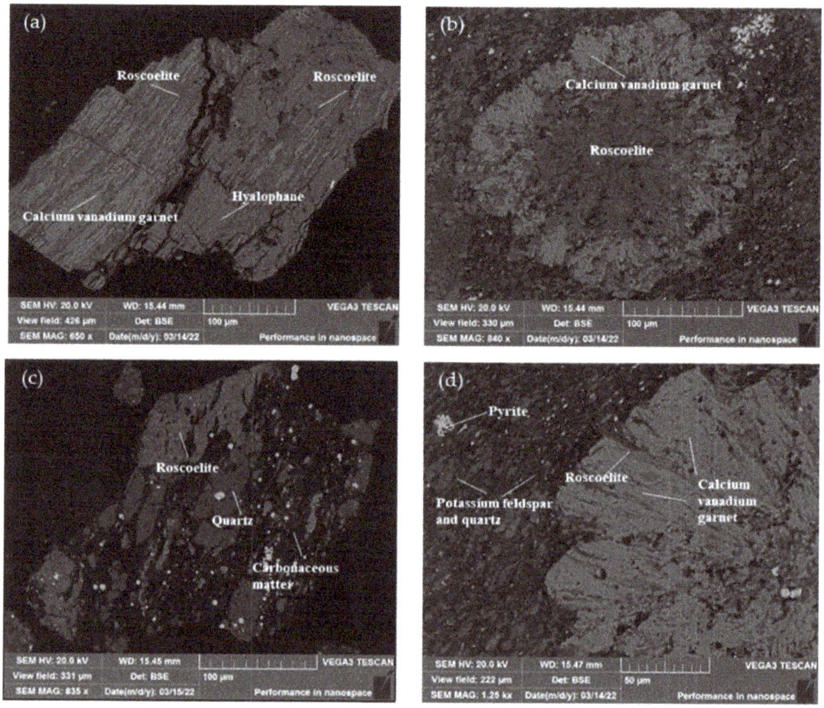

Figure 5. Scanning electron microscope images of roscoelite ($K\{V_2[AlSi_3O_{10}]\}(OH)_2$) in the stone coal: (**a**) picture of roscoelite associated with calcium vanadium garnet and hyalophane ($(K,Ba)Al(Al,Si)_3O_8$); (**b**) picture of roscoelite associated with calcium vanadium garnet; (**c**) picture of roscoelite associated with quartz and carbonaceous matter; (**d**) picture of roscoelite associated with calcium–vanadium garnet and other gangue minerals.

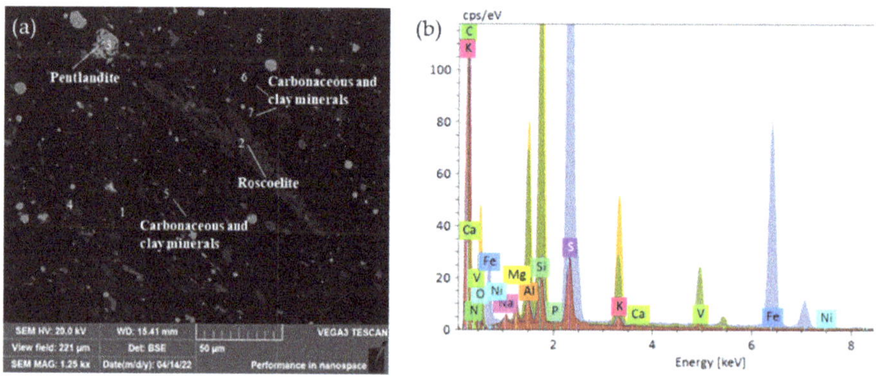

Figure 6. Vanadium occurrence in the ore: (**a**) Scanning electron microscope image of roscoelite and carbonaceous clay minerals in stone coal, 1–8 were the analysis positions of the energy spectral; (**b**) EDS analysis of the minerals containing vanadium.

Table 5. Spectral microdomain composition analysis of vanadium mica and carbonaceous clay minerals at positions 1–8 in Figure 6a.

Number	C	N	O	Na	Mg	Al	Si	P	S	K	Ca	V	Fe	Ni
1	79.05	10.88	5.35	0.71		0.08	0.08		3.36	0.12		0.11	0.25	
2			32.82	0.58	1.56	14.12	27.69		0.41	8.68	0.81	13.33		
3									52.63				45.58	1.79
4			40.51			13.04	33.00			13.45				
5	83.16		4.97	0.41	0.38	1.27	2.78	0.37	3.66	1.10	0.84	0.51	0.54	
6	85.90		4.20	0.41	0.35	1.21	2.61		3.35	1.18		0.51	0.27	
7	89.21		3.64	0.62	0.32	0.53	1.09		3.66	0.43		0.33	0.17	
8	86.88		5.64	0.62	0.50	0.84	1.52		2.94	0.50		0.42	0.14	
Average	84.84	10.88	13.88	0.56	0.62	4.44	9.82	0.37	10.00	3.64	0.83	2.53	7.82	1.79

In summary, the vanadium-containing minerals in stone coal were mainly vanadium mica and vanadium-containing carbonaceous clay minerals. The above vanadium-containing minerals can be pre-concentrated using a beneficiation process, and then the vanadium can be extracted using different leaching methods to reduce the amount of leaching ore and the leaching cost.

(3) Nickel-containing minerals

The nickel-bearing minerals in the stone coal are shown in Table 6. Nickel in the stone coal samples mainly existed in the form of nickel spinel. The nickel-bearing minerals were mainly nickel magnesium spinel and nickel-bearing potassium magnesium alum, accounting for 2.28% and 2.06% of the total mineral content, respectively. The mass fractions of the nickel were 13.43% and 7.05%, respectively. Figure 7 shows the main morphology and dissemination relationship in the nickel-bearing minerals. The Ni-bearing spinel was mostly short-vein aggregate, symbiotic with soluble sodium alum and nickel–vanadium–magnesium ore. The Ni-vanadium-magnesium ore was mainly slender, with a particle size of 10–50 μm, and was embedded around the nickel-bearing spinel.

Table 6. Main nickel minerals in raw ore.

Phase	Proportion of Raw Ore/%
Nickel–magnesium spinel ($(Ni,Mg,Fe)Al_2O_4$)	2.28
Nickel–potassium–magnesium alum ($(Ni,K_2Mg_2[SO_4]_3)$)	2.06
Ferric nickel–magnesium spinel ($((Ni,Fe)Mg_2O_4)$)	0.71
Nickel–ferric–magnesium–aluminum spinel ($((Ni,Mg,Fe)Al_2O_4)$)	0.16
Iron–nickel and vanadium ore $Ni_xFe_y(VO_4)_znH_2O$	0.27
Ferric nickel–aluminum spinel ($((Fe,Ni)Al_2O_4)$)	0.25
Nickel–chromite spinel ($((Fe,Mg)(Al,Cr,Fe)_2O_4)$)	0.08
Trevorite ($NiFe_2O_4$)	0.1
Kolovratite ($Ni_x(VO_4)_z$)	0.03
Nickel and vanadium magnesite ($(Ni,K_2Mg_2(H_2O)_{16}[V_{10}O_{28}])$)	0.04
Nickel–magnesium silicate ($Ni_2[SiO_4]$)	0.01
Total	5.99

(4) Molybdenum-containing minerals

The Mo grade in the stone coal samples was 0.11%. The molybdenum minerals were mainly made up of molybdenite, and the total mineral content accounted for only 0.06%. Figure 8 shows the occurrence states of molybdenite in the stone coal. Combined with the screening results in Table 3, it can be seen that most of the molybdenum was distributed in fine particles. The particle size was generally <5 μm. From the intercalation relationship in the molybdenite listed in Figure 4, it can be seen that the molybdenite was mostly liberated in the form of monomer particles, and some were embedded in carbonaceous materials, carbonaceous clay minerals, quartz, silicon, and humite in the form of fine needles.

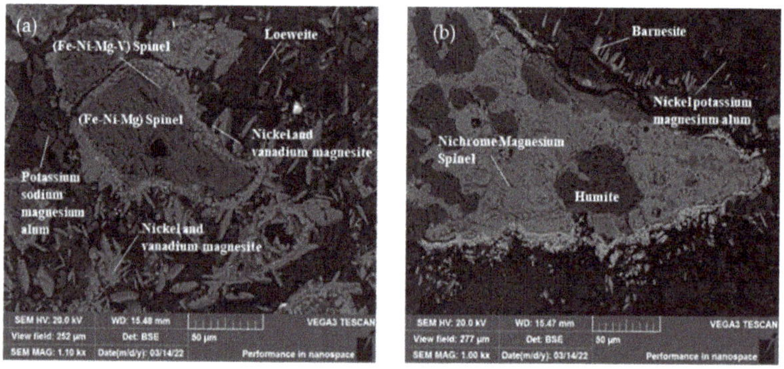

Figure 7. Nickel occurrence in the ore: (**a**) Scanning electron microscope image of nickel spinel in ore; (**b**) Scanning electron microscope image of minerals associated with nickel spinel.

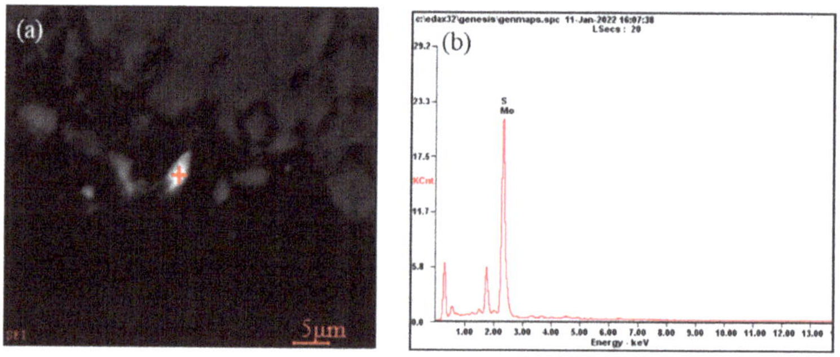

Figure 8. Molybdenite occurrence in the ore: (**a**) Scanning electron microscope image of molybdenite in stone coal; (**b**) EDS analysis for the molybdenite at "+" position.

(5) Soluble minerals

During the screening process, we found that part of the stone coal sample had dissolved in water, resulting in quality loss. Combined with the results of the BPMA, we speculated that the dissolved part was made up of soluble (sodium, magnesium, and potassium) sulfate minerals, which accounted for about 12.45% of the total mineral amount. To further verify the chemical composition of the soluble substances, a 10 g sample was stirred in 50 mL of deionized water for 1 h. In total, 8.3 g of residue was left after stirring. The main components in the solution are shown in Table 7. Part of the filtrate was evaporated and crystallized, and the crystalline sample was analyzed using SEM-EDS and XRD. The results are shown in Figure 9.

Table 7. Content of the main elements before and after dissolution.

Elements	K	Ca	Na	Mg	V	Ni	Mo
Filter residue (%)	1.34	2.19	0.7	5.7	0.75	0.48	0.11
Filtrate (g/L)	0.23	0.38	4.01	1.39	0.05	0.09	0.04
Raw ore (%)	1.27	2.16	3.72	5.8	0.68	0.51	0.12

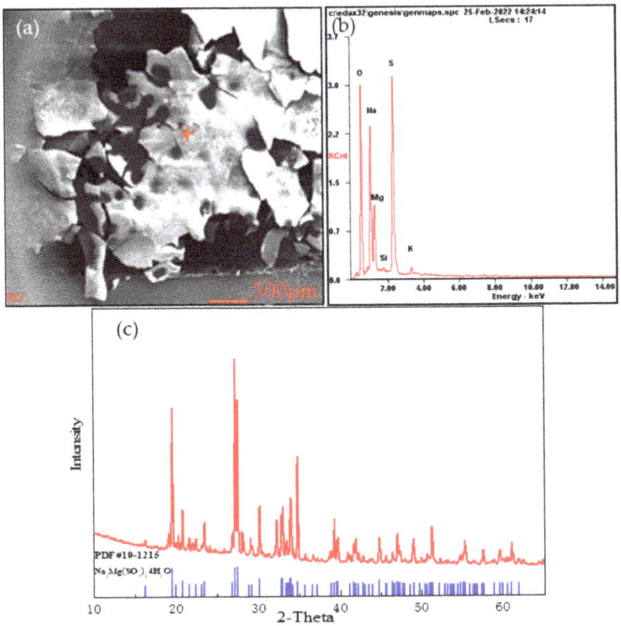

Figure 9. Soluble minerals: (**a**) Scanning electron microscope image of the soluble minerals; (**b**) EDS analysis at "+" position; (**c**) X-ray diffraction analysis of the soluble minerals.

According to the changes in the main elements before and after dissolution and an analysis of the crystal spectrum, the soluble parts of the stone coal sample were found to be sodium–magnesium alum, potassium–sodium–magnesium alum, and nickel–sodium–magnesium alum. In addition, a small amount of V, Mo, and Ni was also detected in the filtrate.

3.2. Principle Process Determination

The V_2O_5, Ni, and Mo contents in the stone coal samples were 1.29%, 0.53%, and 0.11%, respectively, and the organic carbon content was 6.13%. Vanadium mainly occurred in the vanadium mica and carbonaceous clay minerals; nickel mostly existed in the form of spinel; molybdenum existed in the form of molybdenite with a high degree of dissociation. Vanadium in the form of carbonaceous clay minerals can be recovered using scrubbing and desliming techniques [24–26]. Vanadium mica can be recovered using flotation [27–29]. Liberated molybdenite has good floatability and can be enriched using flotation [30]. In addition, some molybdenite and carbonaceous clay minerals are associated closely, so scrubbing desliming can simultaneously enrich molybdenum and vanadium. Because the mineral properties of nickel-bearing spinel and vanadium-bearing minerals are quite different, the beneficiation process cannot be used for simultaneous enrichment, but vanadium and nickel can be extracted with leaching.

The mud content in the stone coal samples is high, and the mud will adhere to the surface of useful minerals and gangue minerals, which will affect the flotation reagents on the surface of the value minerals. Thus, desliming is also required before flotation [31]. Crushed stone coal samples have fine ore particle sizes and can be deslimed without grinding.

The difficulty of using the flotation process for vanadium mica is separating the vanadium mica from gangue minerals such as quartz and feldspar. Under alkaline conditions, collectors such as sodium oleate [32,33], sodium petroleum sulfonate [34], and dodecyl ammonium chloride [35] can be used to separate non-metallic minerals from gangue minerals such as quartz. Studies have shown that the zero point of mica is very low, and fine mica can be recovered by using cationic collectors in an acidic medium [36,37]. However, due to the complex solution ions in the pulp and the fact that quartz, feldspar, and mica are silicate minerals, it is necessary to enrich vanadium mica with highly efficient and selective collectors [38,39].

Based on the above analysis, combined with the mineralogical characteristics of stone coal samples, we propose a combined process for vanadium, nickel, and molybdenum recovery, as shown in Figure 10.

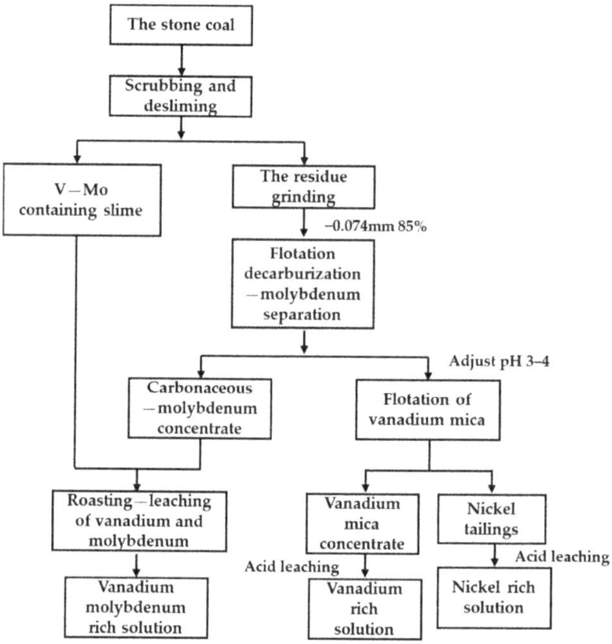

Figure 10. Basic process flow chart of vanadium, nickel, and molybdenum recovery.

The crushed stone coal sample (−2 mm) contained many fine particles, which could be directly scrubbed and deslimed using sodium silicate as a dispersant. After desliming, the tailings were reground to 85%—0.074 mm, and the minerals containing molybdenum, vanadium, and nickel were fully liberated before flotation [40–42]. Carbonaceous matter in stone coal interferes with other minerals during flotation, so decarburization was required before molybdenum and vanadium flotation could proceed. The collector used for flotation–decarbonization was kerosene, and the collectors used for molybdenum flotation were butylammonium dithiophosphate and butyl xanthate. Then, the pH of the pulp was adjusted to 3~4, and the vanadium mica was collected by 3030C. Vanadium and molybdenum needed to be extracted with a roasting–leaching process due to the high carbon content in the vanadium–molybdenum slime obtained by scrubbing and the molybdenum concentrate obtained using flotation, while the vanadium mica and nickel-bearing minerals enriched by flotation could be directly acid-leached under mild conditions.

3.3. *Exploratory Experiments Based on Mineralogy*

3.3.1. Beneficiation Enrichment Experiments

Combined with the mineralogical features study, the scrubbing–desliming–flotation method was selected to enrich vanadium, molybdenum, and nickel.

First, the raw ore was scrubbed and deslimed. At a certain pulp concentration, sodium silicate was added to disperse the mineral particles. Then, the pulp was stirred at high strength. Finally, the fine mud vanadium concentrate was obtained via sedimentation and classification. The coarse tailings were separated via grinding and flotation, and the vanadium and nickel were separated from the coarse scrubbing tailings via flotation. The test results are shown in Table 8, and the process flow with the assay of each product is shown in Figure 11.

Table 8. Experimental results for scrubbing–desliming–flotation.

Sample	Yield(x ± σ)/%	Grade/%			Recovery (x ± σ)/%		
		V_2O_5	Mo	Ni	V_2O_5	Mo	Ni
V-Mo Slime	15.29 ± 0.00	4.10	0.44	0.76	41.31 ± 0.47	51.40 ± 0.14	17.64 ± 0.22
Carbonaceous matter	9.78 ± 0.61	1.19	0.19	0.31	7.68 ± 0.81	14.21 ± 0.63	4.61 ± 0.16
Molybdenum concentrate	5.01 ± 0.6	1.27	0.47	0.23	4.20 ± 0.45	18.01 ± 0.54	1.78 ± 0.18
Vanadium Mica	27.26 ± 1.70	1.52	0.07	0.55	27.31 ± 1.77	14.58 ± 0.36	22.85 ± 2.56
Tailings	23.48 ± 1.29	1.26	0.01	1.49	19.50 ± 1.67	1.79 ± 0.31	53.13 ± 2.03
Soluble fraction	19.18	-	-	-	-	-	-
Raw ore	100.00	1.49	0.13	0.59	100.00	100.00	100.00

Note: x—mean value, σ—confidence interval. Number of samples is 3.

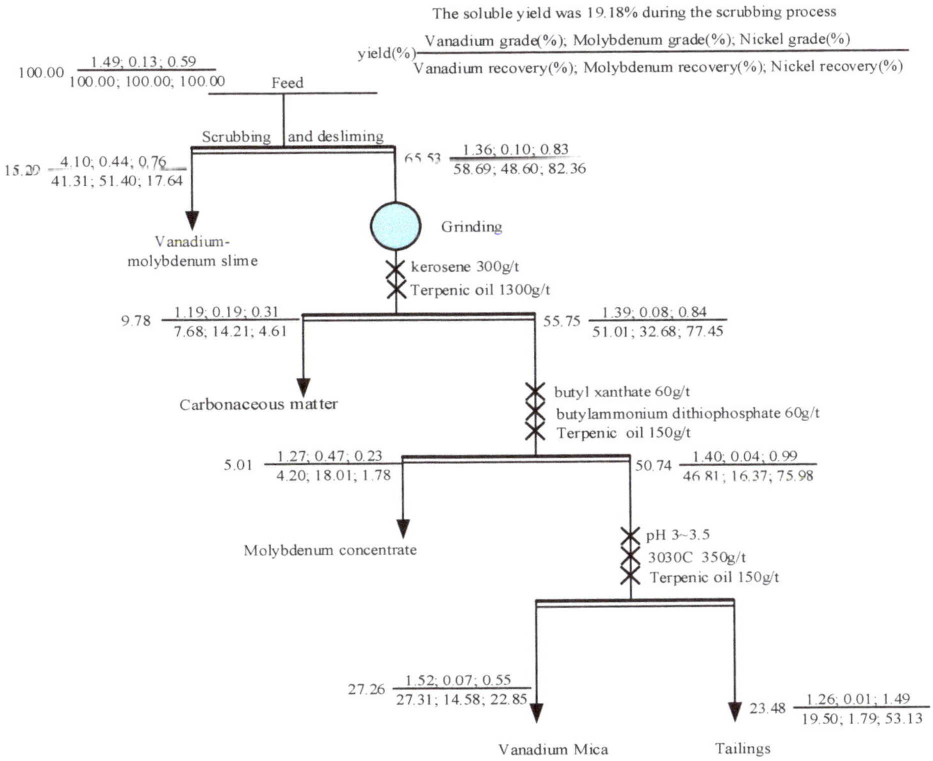

Figure 11. Flow chart of quantity and quality.

From the results in Table 8, it can be seen that scrubbing–desliming and flotation enriched the vanadium, nickel, and molybdenum in different products. The grade of V_2O_5 in the vanadium-molybdenum slime was 4.10%, and the recovery was 41.31%. The Mo grade was 0.44% and the recovery was 51.40%. Based on the mineralogical characteristics of molybdenum, a small amount of molybdenum was enriched in the carbonaceous matter and molybdenum concentrate, but the recovery was not high. Based on a phase analysis of the vanadium, we found that, in addition to clay minerals, it was also partially present in mica. However, the enrichment effect of vanadium mica in the flotation test results was poor, and the V_2O_5 grade in the concentrate was only 1.52%. This is because the stone coal contained many gangue minerals with a similar floatability to vanadium-containing mica, such as silicate and quartz, which made the flotation separation process difficult. Furthermore, the vanadium was partially substituted in the mica, so the mica contained less vanadium, and the flotation process produced vanadium-containing mica with similar mineral properties to vanadium-free aluminosilicate, resulting in more concentrate and yielding small improvements in its grade [10,43]. Based on the occurrence states of nickel in the stone coal, it can be seen that the floatability of nickel was poor, the nickel grade in the tailings was 1.49%, and the recovery was 53.13%. Because stone coal also contained nickel–vanadium symbiotic ore, a small part of the nickel was produced in the vanadium-containing mica concentrate. The above results demonstrate that this process can effectively recover vanadium, nickel, and molybdenum from stone coal by multiple steps.

3.3.2. Vanadium, Nickel, and Molybdenum Leaching Experiments

Corresponding leaching experiments were carried out for different products obtained in the beneficiation experiments. Tang [44] and Zhao [45]'s experiments showed that carbon can be completely removed by roasting stone coal at 700 °C for 1 h. In our laboratory work, the vanadium–molybdenum slime, carbonaceous matter, and molybdenum concentrate were mixed and roasted at 650 °C for 2 h. The roasted slag was leached for 10 h at room temperature, a sulfuric acid concentration of 1.84 mol/L, and a liquid–solid ratio of 4:1. The vanadium-containing mica concentrate and nickel-containing tailings were leached under the same condition as above for the roasted slag. The results are shown in Table 9. At the same time, a fluoride leaching experiment was carried out on the raw ore. The results are shown in Table 10.

Table 9. Leaching efficiency of different beneficiation products.

Sample	Leaching Efficiency (x ± σ)/%		
	V	Ni	Mo
Slime-Carbon-Molybdenum Concentrate	72.55 ± 2.63	58.28 ± 2.15	94.10 ± 0.16
Vanadium mica	50.74 ± 0.27	71.67 ± 0.71	3.25 ± 0.80
Nickel-bearing tailings	72.45 ± 1.60	67.86 ± 1.02	13.58 ± 1.58

Note: x—mean value, σ—confidence interval. Number of samples is 3.

Table 10. Experimental conditions and results of fluorination leaching in the stone coal raw ore.

Elements	V	Ni	Mo
Leaching efficiency/%	54.89	67.35	24.66

Leaching conditions. Sulfuric acid concentration 2.5 mol/L; Calcium fluoride 10%; Liquid–solid ratio 1.5:1; Leaching time 8 h.

It can be seen from Table 10 that the leaching efficiency of the molybdenum in the slime–carbon–molybdenum concentrate was as high as 94.10% after roasting. This was due to the oxidation of MoS_2 to MoO_3 at 650 °C, which is more easily leached using sulfuric acid. The leaching efficiency of the vanadium was 72.55%. As vanadium in clay minerals mostly exists in the form of V(IV) and V(V) adsorption states, it can be dissolved in acid [10]. The leaching efficiency of the vanadium in the vanadium-containing mica was low, which may be related to the crystal structure of the vanadium mica. Therefore, in the process of leaching, it is necessary to add leaching aids to strengthen the fracturing of Si-O and Al/(V)-O in mica [46–48]. According to the results of the direct fluorination leaching of the raw ore in Table 10, it can be seen that the vanadium, nickel, and molybdenum were first pre-concentrated by the beneficiation process, after which the roasting–sulfuric-acid-leaching or direct acid leaching processes were used for different products to reduce acid consumption without producing fluorine-containing wastewater and reducing pollution in water resources.

4. Conclusions

(1) V_2O_5, Ni, and Mo grades in the stone coal were 1.29%, 0.53%, and 0.11%, respectively. The vanadium-bearing minerals were mainly carbonaceous clay minerals and vanadium-bearing mica. The nickel-bearing minerals were mainly nickel magnesium spinel with a small amount of nickel-bearing potassium–magnesium alum. The molybdenum-bearing minerals were mainly molybdenite mostly in the form of molybdenite monomer and partially embedded in carbonaceous materials, carbonaceous clay, and other minerals. The stone coal sample also contained soluble minerals such as sodium magnesium alum, potassium sodium magnesium alum, and nickel–sodium–magnesium alum. The loss of vanadium, nickel, and molybdenum in soluble fractions was comparatively small.

(2) Based on the research results of the mineralogical features study, we propose enriching vanadium, nickel, and molybdenum using a scrubbing–desliming–flotation process to obtain vanadium-rich molybdenum slime and nickel-containing tailings. In our experiment, the V_2O_5 grade increased from 1.49% to 4.10%, and the recovery was 41.31%. The Mo grade increased from 0.13% to 0.44%, and the recovery was 52.70%. The Ni grade in the flotation tailings was 1.49%.

(3) The obtained vanadium-molybdenum slime, carbonaceous matter, and molybdenum concentrate were roasted and acid-leached; the leaching efficiency of V was 72.55% and the leaching efficiency of Mo was 94.10%. The vanadium-bearing mica concentrate and tailings were directly leached using acid. When the dosage of sulfuric acid was 1.84 mol/L, the leaching efficiency of V and Ni were 50.74% and 67.86%, respectively. Compared with the direct fluorination leaching of the raw ore, the leaching effects of vanadium and nickel were similar, but the leaching efficiency of molybdenum was higher with a lower dosage of sulfuric acid. The current test was only an exploration and needs to be optimized in the follow-up work.

(4) After a mineralogical features study of stone coal from Shaanxi Province, the phase composition and inlay characteristics of vanadium, nickel, and molybdenum were identified, and a scrubbing and sludge–flotation–leaching process was selected to realize the green and clean enrichment of vanadium, nickel, and molybdenum metals. This may provide guidance and reference for industrial applications.

Author Contributions: Conceptualization, M.W. and L.C.; methodology, M.W.; software, L.C.; validation, W.L., J.W. and H.Y.; formal analysis, M.W. and X.Y.; investigation, M.W.; resources, J.W.; writing—original draft preparation, M.W.; writing—review and editing, W.L. and J.W.; visualization, L.C.; supervision, J.W.; project administration, J.W., X.Y. and W.L.; funding acquisition, J.W. and X.Y. All authors have read and agreed to the published version of the manuscript.

Funding: This research was funded by the National Key Research and Development Project (2020YFC1909703).

Data Availability Statement: The data that support the findings of this study are available from the corresponding author upon reasonable request.

Acknowledgments: The authors are grateful to the financial support from the National Key Research and Development Project (2020YFC1909703).

Conflicts of Interest: The authors declare no conflict of interest.

References

1. Santos, D.A.; Dixit, M.K.; Kumar, P.P.; Banerjee, S. Assessing the role of vanadium technologies in decarbonizing hard-to-abate sectors and enabling the energy transition. *iScience* **2021**, *24*, 103277. [CrossRef] [PubMed]
2. Wang, X.; Wang, A.; Zhong, W.; Zhu, D.; Wang, C. Analysis of international nickel flow based on the industrial chain. *Resour. Policy* **2022**, *77*, 102729. [CrossRef]
3. Liu, B.; Zhang, B.; Han, G.; Wang, M.; Huang, Y.; Su, S.; Xue, Y.; Wang, Y. Clean separation and purification for strategic metals of molybdenum and rhenium from minerals and waste alloy scraps—A review. *Resour. Conserv. Recy.* **2022**, *181*, 106232. [CrossRef]
4. Vind, J.; Tamm, K. Review of the extraction of key metallic values from black shales in relation to their geological and mineralogical properties. *Miner. Eng.* **2021**, *174*, 107271. [CrossRef]
5. Kenzhaliyev, B.K.; Surkova, T.Y.; Azlan, M.N.; Yulusov, S.B.; Sukurov, B.M.; Yessimova, D.M. Black shale ore of Big Karatau is a raw material source of rare and rare earth elements. *Hydrometallurgy* **2021**, *205*, 105733. [CrossRef]
6. Dai, S.; Zheng, X.; Wang, X.; Finkelman, R.B.; Jiang, Y.; Ren, D.; Yan, X.; Zhou, Y. Stone coal in China: A review. *Int. Geol. Rev.* **2018**, *60*, 736–753. [CrossRef]
7. Dai, Z.; Zhang, E. Technical status of leaching vanadium from stone coal vanadium ore. *Min. Metall. Eng.* **2015**, *35*, 85–88.
8. Bao, S.; Chen, B.; Zhang, Y. Research status and prospect of vanadium extraction technology for vanadium-bearing shale in China. *Metal. Mine.* **2020**, *10*, 20.
9. Kang, X.; Ye, G.; Zhu, S.; Liang, X. Research progress on V_2O_5 extraction from vanadium-bearing shale. *Iron. Steel Vanadium. Titan.* **2022**, *43*, 25–34.
10. Zhang, C.; Niu, M.; Chu, F. Analysis on occurrence characteristics and extraction process of vanadium in stone coal. *Mod. Min.* **2018**, *34*, 91–95.
11. Wang, M.; Huang, S.; Chen, B.; Wang, X. A review of processing technologies for vanadium extraction from stone coal. *Min. Proc. Ext. Met.* **2020**, *129*, 290–298. [CrossRef]
12. Li, M.; Liang, D.; He, X. Study on mineralogy of a stone-coal vanadium ore from Hubei. *Metal Mine.* **2015**, *2*, 87–91.
13. Zhang, W.; Hou, E.; Yang, J.; Li, H.; Huan, B. Study on vanadium-molybdenum-selenium and other associated elements in stone coal. *Rare Metals* **2019**, *43*, 1092–1102.
14. Li, Y. Geochemistry characteristics of Ni Mo Pt group elements in the lower Cambrian bone coal in Hunan Province. *J. China Coal Soc.* **1996**, *3*, 38–41.
15. Wang, L.; Sun, W.; Zhang, Q. Recovery of vanadium and carbon from low-grade stone coal by flotation. *T. Nonferr. Metal. Soc.* **2015**, *25*, 3767–3773. [CrossRef]
16. Liu, C.; Zhang, Y.; BAO, S. Vanadium recovery from stone coal through roasting and flotation. *T. Nonferr. Metal. Soc.* **2017**, *27*, 197–203. [CrossRef]
17. Gu, X.; Sun, W.; Liu, R.; Song, S.; Chen, X. Study on mineral processing of a decarburized stone coal in Hunan. *Nonferrous Met. Miner. Process. Sect.* **2014**, *5*, 67–71.
18. Liu, Q. *Flotation and Vertical Roasting Process for Extracting Molybdenum from Vanadium-Bearing Molybdenum Coal Mine*; Hubei Qinjiahe Mining Co., Ltd.: Hubei, China, 2014.
19. Zhou, J.; Gu, Y. Chapter-6—Geometallurgical Characterization and Automated Mineralogy of Gold Ores. In *Gold Ore Process*, 2nd ed.; Elsevier: Amsterdam, The Netherlands, 2016; pp. 95–111.
20. Chen, B.; Ma, L.; Chen, Y.; Ma, H. Chemical phase of vanadium in stone coal. *Phys. Test. Chem. Anal. Part B Chem. Anal.* **2017**, *53*, 1052–1056.
21. Wen, L.; Jia, M.; Wang, Q.; Fu, Q.; Zhao, J. A new SEM-based automated mineralogy system: BPMA and its application prospects in mining industry. *Nonferrous Met. Miner. Process. Sect.* **2021**, *2*, 12–23.
22. Wen, L.; Fu, Q.; Yu, Z.; Jia, M. Application of Automated Quantitative Mineralogy System in the Process Mineralogy Study of Low-grade Fine-grained Molybdenum Ore. *Nonferrous Met. Miner. Process. Sect.* **2022**, *2*, 31–38.
23. Wang, L.; Zhang, Q.; Sun, W. Research on Occurrence of Vanadium in Stone Coal Deposit at Loufanggou Area, Shangluo City, Shanxi Province, China. *Acta Mineral. Sin.* **2017**, *37*, 29–35.
24. Liu, J.; Sun, W.; Wang, L.; Liu, W.; Zhang, M. Study on a New Beneficiation Technology of a Vanadium-Bearing Stone Coal Ore in Shaanxi. *Nonferrous Met. Miner. Process. Sect.* **2015**, *2*, 58–63.
25. Chen, C.; Zuo, H.; Wang, H. Research on vanadium ore dressing technology. *China Met. Bull.* **2018**, *11*, 230.
26. Liu, X.; Zhang, Y.; Bao, S.; Ren, L.; Liu, C. Experimental Research on Selective Desliming of Vanadium Bearing Stone Coal Ore. *Nonferrous Met. Miner. Process. Sect.* **2015**, *6*, 41–44.
27. Wang, L.; Sun, W.; Liu, R.; Gu, X. Flotation recovery of vanadium from low-grade stone coal. *T. Nonferr. Metal. Soc.* **2014**, *24*, 1145–1151. [CrossRef]
28. Sun, W.; Gu, X.; Liu, R.; Ma, Y.; Zheng, B. Occurrence State of Vanadium Minerals in Carbonaceous Stone Coal and Corresponding Beneficiation Technology. *Min. Metall. Eng.* **2013**, *33*, 28–31.
29. Xing, X.; Wan, H.; Ning, S.; She, Z. Experimental Research on Mineral Processing and Vanadium Extraction from a Stone Coal Vanadium Ore. *Nonferrous Met. Miner. Process. Sect.* **2016**, *3*, 43–47.
30. Yi, G.; Macha, E.; Dyke, J.; Ed Macha, R.; McKay, T.; Free, M.L. Recent progress on research of molybdenite flotation: A review. *Adv. Colloid Interface Sci.* **2021**, *295*, 102466. [CrossRef]

31. Chen, X.; Yang, J.; Mao, Y.; Liu, H. An Innovation Process for Comprehensive Utilization of Low-Grade Clay-Vanadium Ores. *Nonferrous Met. Miner. Process. Sect.* **2010**, *5*, 9–12.
32. Jin, J.; Gao, H.; Ren, Z.; Chen, Z. The flotation of kyanite and sillimanite with sodium oleate as the collector. *Minerals* **2016**, *6*, 90. [CrossRef]
33. Ren, Z.; Shen, Y.; Gao, H.; Chen, H.; Liu, C.; Chen, Z. Comparison of Sodium Oleate and Sodium Petroleum Sulfonate for Low-Temperature Flotation of Fluorite and the Collecting Mechanisms. *Min. Metall. Explor.* **2021**, *38*, 2527–2536. [CrossRef]
34. Chen, Z.; Ren, Z.; Gao, H.; Zheng, R.; Jin, Y.; Niu, C. Flotation studies of fluorite and barite with sodium petroleum sulfonate and sodium hexametaphosphate. *J. Mater. Res. Technol.* **2019**, *8*, 1267–1273. [CrossRef]
35. Chen, Z.; Ren, Z.; Gao, H.; Lu, J.; Jin, J.; Min, F. The effects of calcium ions on the flotation of sillimanite using dodecylammonium chloride. *Minerals* **2017**, *7*, 28. [CrossRef]
36. Di, Y.; Jiang, A.; Huang, H.; Luo, Q.; Wei, W.; Wang, R.; Chen, S. Molecular dynamics simulations of adsorption behavior of DDAH, NaOL and mixed DDAH/NaOL surfactants on muscovite (001) surface in aqueous solution. *J. Mol. Graphics Modell.* **2022**, *113*, 108161. [CrossRef] [PubMed]
37. Chen, F.; He, D.; Deng, B.; Hu, Y.; Wang, Q. Flotation Separation Behavior of Muscovite and Quartz. *Metal. Mine.* **2019**, *8*, 108–112.
38. Ren, L.; Qiu, H.; Zhang, Y.; Nguyen, A.V.; Zhang, M.; Wei, P.; Long, Q. Effects of alkyl ether amine and calcium ions on fine quartz flotation and its guidance for upgrading vanadium from stone coal. *Powder Technol.* **2018**, *338*, 180–189. [CrossRef]
39. Zou, Z.; Luo, H.; Tang, J. Effect of common Metal Ions on the Flotation Performance of Muscovite and Quartz. *Conserv. Util. Miner. Resour.* **2018**, *1*, 66–71.
40. Lv, H.; Xie, X.; Tong, X.; Chen, L.; Chen, Y. Flotation of carbon from a vanadium leaching residue of unroasted stone coal. *Int. J. Oil Gas Coal Technol.* **2018**, *19*, 217–229. [CrossRef]
41. Wang, C.; Yang, Y. Flotation Technology for Reducing Carbon Content in Vanadium-bearing Stone Coal. *Min. Metall. Eng.* **2018**, *38*, 64–66. [CrossRef]
42. Zhang, L. Study on Mineral Processing of a Low-Grade Vanadium Stone Coal Containing High Grade Carbon. *Hunan Nonferrous Met.* **2015**, *31*, 10–14.
43. Ye, G.; Zhu, S.; Chen, Z.; Hu, Y.; Dong, X.; Zhang, H. A Research Review on Beneficiation Pre-Concentration of Vanadium-Bearing Stone Coal. *Chin. J. Rare Met.* **2022**, *46*, 120–130.
44. Tang, J.; Zhang, Y.; Bao, S.; Liu, C. Pre-concentration of vanadium-bearing mica from stone coal by roasting-flotation. *Physicochem. Probl. Miner. Process.* **2017**, *53*, 402–412.
45. Zhang, J.; Zhang, Y.; Huang, J.; Liu, T. Process of Blank Roasting-Sulphuric Acid Leaching of Vanadium with Leaching Agent from Stone Coal. *Rare Met.* **2013**, *37*, 446–452.
46. Huang, Z.; Chen, T.; Lin, H.; Zhou, Y.; Yan, B.; Tian, H. Research Progress in Vanadium Recovery by Hydrometallurgical Acid Leaching from Stone Coal. *Rare Met. Cem. Carbides* **2020**, *3*, 11–17.
47. Zhang, Y.; Xue, N.; Liu, T.; Huang, J. Green extraction technology of vanadium-bearing shale via whole hydrometallurgical process. *J. China U. Min. Techno.* **2022**, *3*, 520–531.
48. Zheng, Q.; Zhang, Y.; Liu, T.; Huang, J.; Xue, N. Removal Process of Structural Oxygen from Tetrahedrons in Muscovite during Acid Leaching of Vanadium-Bearing Shale. *Minerals* **2018**, *8*, 208. [CrossRef]

minerals

MDPI

Article

Selective Adsorption of Sodium Silicate on the Surface of Bastnaesite and Fluorite in Salicylhydroxamic Acid System under Alkaline Conditions

Maoyuan Wang [1], Wenliang Xiong [1,2,3,*], Junhui Xiao [4], Yao Guo [1,2,3], Jie Deng [1,2,3], Da Chen [1,2,3], Anni Ouyang [1], Menglin Lei [1,2,3] and Lijun Zhang [1,2,3,*]

1 Institute of Multipurpose Utilization of Mineral Resources, Chinese Academy of Geological Sciences, Chengdu 610041, China
2 Sichuan Rare Earth Technology Innovation Center, Chengdu 610041, China
3 Engineering and Technology Innovation Center for Comprehensive Utilization of Strategic Minerals, Ministry of Natural Resources, Chengdu 610041, China
4 School of Environment and Resource, Southwest University of Science and Technology, Mianyang 621010, China
* Correspondence: seanwlzm@163.com (W.X.); mangran0627@163.com (L.Z.)

Abstract: During the flotation separation process of bastnaesite, it is difficult to separate bastnaesite from fluorite effectively. In this present study, sodium silicate (SS) can effectively improve the flotation separation effect of bastnaesite and fluorite in salicylhydroxamic acid (SHA) systemasa. Through relevant analyses, such as Zeta potential measurements, adsorption capacity tests, Fourier transform infrared (FTIR) spectroscopic analyses and X-ray photoelectron spectroscopy (XPS) tests, the selective suppressor of SS on fluorite was proven. At pH 10, the single mineral flotation results show that with the increase of SS dosage, the flotation recovery of fluorite rapidly decreases from 61.5% to 35.31%, while the flotation rate of bastnaesite is still high (recovery is 80.02%). Then, the experiment of artificial mixed ore proved that the flotation separation of fluorite and bastnaesite was effective under the appropriate dosage of inhibitor. The results of potentiodynamic measurement and an adsorption capacity test showed that the $SiO(OH)_3^-$ structure of SS more easily reacted with fluorite, which further prevented the adsorption of SHA on the fluorite surface. FTIR test results and XPS analysis further showed that SS had a strong binding effect with the Ca site on the fluorite surface, but a weak binding effect with the Ce site on the bastnaesite surface. Consequently, SS can be used as an effective inhibitor in the flotation separation of fluorite and bastnaesite.

Keywords: bastnaesite; fluorite; flotation; sodium silicate; salicylhydroxamic acid

Citation: Wang, M.; Xiong, W.; Xiao, J.; Guo, Y.; Deng, J.; Chen, D.; Ouyang, A.; Lei, M.; Zhang, L. Selective Adsorption of Sodium Silicate on the Surface of Bastnaesite and Fluorite in Salicylhydroxamic Acid System under Alkaline Conditions. *Minerals* **2023**, *13*, 69. https://doi.org/10.3390/min13010069

Academic Editor: Chiharu Tokoro

Received: 23 November 2022
Revised: 20 December 2022
Accepted: 27 December 2022
Published: 31 December 2022

1. Introduction

The 17 metals in the periodic table (lanthanides, scandium and yttrium) make up the rare-earth elements, which include light, medium and heavy REEs [1–3]. Due to their unique physical and chemical properties, light REEs are widely used in catalysis, superconductivity, the chemical industry and other fields. Light REEs are widely used in many fields due to their special characteristics [4]. Today, demand for light REEs is surging in many countries around the world [5,6]. China has a large amount of rare-earth resources, with about 65.88 million tons of rare-earth reserves, and the country supplies more than 90% of the world's rare-earth resources [7,8]. Normally, bastnaesite is associated with calcite, barite, fluorite, etc. How to effectively enrich bastnaesite from gangue minerals is of great importance for the development of low-grade fine bastnaesite [9,10]. Flotation separation is a highly effective method to recover slight bastnaesite, as compared to magnetic and gravity separation. Bastnaesite and fluorite have similar properties. The effective separation of fine bastnaesite from fluorite has always been a difficult problem in the field of mineral processing [11–14]. Salicylhydroxamic acid (SHA) is highly selective to bastnaesite and is usually used as a collector in the bastnaesite flotation process. However, SHA is also selective for fluorite [15–17]. Therefore, the search for safe, effective and cost-efficient inhibitors can increase the separation efficiency of bastnaesite and fluorite. The inhibitors can provide preferential adsorption on gangue minerals due to their increased hydrophilicity and effectively prevent adsorption between collectors and mineral surface sites [18–20].

Sodium silicate (SS) is the generally used fluorite inhibitor in bastnaesite flotation, and SS has a good dispersion effect on pulp [21]. The adsorption mechanism of SS on mineral surfaces involves physical adsorption and chemical adsorption. With SS, hydrophilic and electronic silica colloidal particles and hydrogen silicate are the main inhibitors [22]. The study found that, in the flotation separation process, the absolute value of the adsorption energy of fluorite for $SiO(OH)_3^-$ was obviously higher than that of $Si(OH)_4$, and the adsorption of $SiO(OH)_3^-$ was more secure on fluorite surfaces than that of $Si(OH)_4$ [23]. It is possible that $SiO(OH)_3^-$ reacted with Ca^{2+} to form a complex, changing its hydrophilicity and, thereby, reducing the adverse effects of fluorite on the flotation of bastnaesite. Under different pH systems, SS displayed inhibitory effects with different intensities on fluorite. The difference in surface hydrophobicity was used to separate fluorite from bastnaesite [14,17].

This study used SS as an inhibitor for flotation separation of bastnaesite and fluorite, while SHA was used as a collector. The inhibition effect of SS on fluorite at a specific pH value was compared by micro-flotation experiments. At the same time, the selective inhibition effect of SS on fluorite was explained by potentiodynamic measurements, adsorption tests, infrared spectroscopy, XPS analysis and solution chemistry. Thus, the conclusion of this study can provide a theoretical basis for the development and implementation of the separation process of fine-grained bastnaesite and co-associated fluorite resources.

2. Materials and Method

2.1. Materials

The pure minerals of bastnaesite and fluorite used in the experiment were sourced from China. The single mineral utilized in the test was obtained by crushing, picking and repeatedly purifying the raw ore. The purity of the bastnaesite and fluorite was determined by chemical analysis (Tables 1 and 2) and X-ray diffraction (Figure 1), which proved that the high purity met the test requirements [24]. The test regulators (NaOH and H_2SO_4), collector (Salicylhydroxamic acid) and inhibitor (Sodium silicate) were all analytical pure.

Table 1. Chemical composition of bastnaesite powder samples (%).

Element	**CeO_2**	**La_2O_3**	**Nd_2O_3**	**Pr_2O_3**	**Gd_2O_3**	**Sm_2O_3**	**Fe_2O_3**
Content	34.882	29.476	7.096	2.624	0.564	0.301	1.348
Element	MoO_3	BaO	PbO	CaO	SiO_2	ThO_2	Others
Content	1.255	1.211	1.114	0.632	0.416	0.320	0.990

Table 2. Chemical composition of fluorite powder samples (%).

Element	**CaO**	**F**	**SiO_2**	**Al_2O_3**	**SO_3**
Content	73.177	23.705	0.172	0.022	0.020
Element	Fe_2O_3	ZnO	SrO	Y_2O_3	Others
Content	0.050	0.020	0.009	0.007	0.011

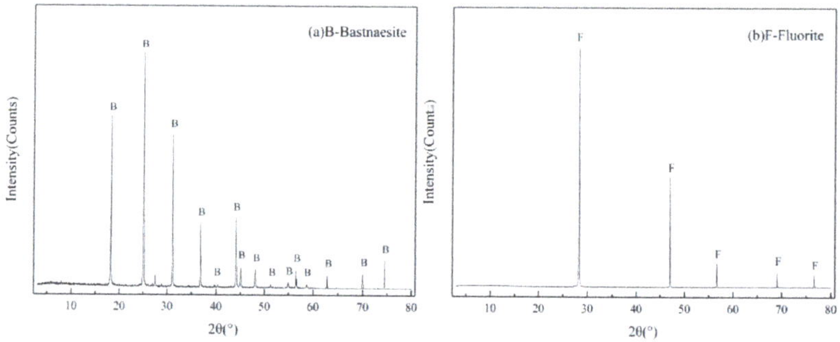

Figure 1. XRD spectra of bastnaesite (**a**) and fluorite (**b**) particles.

2.2. Micro-Flotation Tests

The pure mineral flotation tests used the XFGII hanging trough flotation machine shown in Figure 2. Each 2.00 g sample was placed in a 50 mL flotation tank before adding 35 mL deionized water and stirring for 1 min. The pH was adjusted with H_2SO_4 or NaOH for 3 min. Samples were stirred for 3 min after addition of SS and then for another 3 min with SHA. After adding the foaming agent, stir for 2 min and then flotation for 4 min. The foam products were dried and weighed [25], and the calculation formula of flotation recovery is as follows. The micro-flotation tests flow is shown in Figure 3.

$$\varepsilon = \frac{m_1}{m_1 + m_2} \times 100\% \tag{1}$$

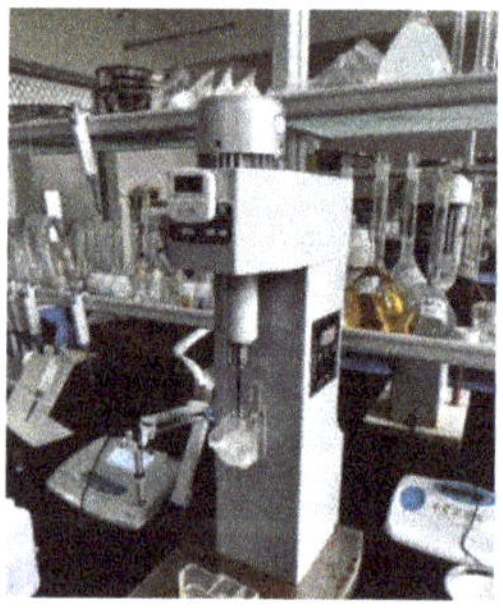

Figure 2. XFG flotation machine used for the flotation tests.

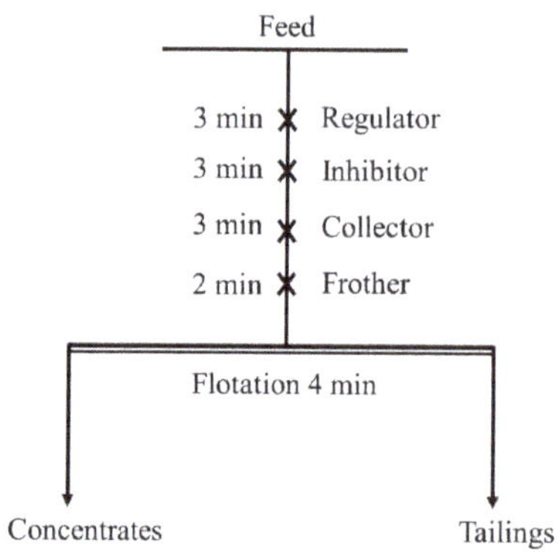

Figure 3. Flotation flow chart for minerals.

2.3. Potentiodynamic Potential Measurement

The Malvern Zeta sizer nano ZS analyzer was used to test mineral Zeta potential. The specific test steps are as follows: take 2.00 mg single mineral (−5 μm) into the conical flask, add 30 mL deionized water into it, and adjust the pulp according to the flotation process for a single mineral. It was oscillated on the thermostatic vibrating screen for 30 min and left for 10 min. The supernatant was taken to measure zeta potential. Three measurements were taken under different dose conditions, and the average of the final results was taken.

2.4. Adsorption Tests

A UV-3000 ultraviolet spectrophotometer was used to measure the adsorption capacity of the reagent on the mineral surface. A 2.00 g sample of a single mineral was weighed and placed in a conical flask for each test, and 35 mL deionized water was added. The reagent was added according to the single mineral flotation process. After shaking in a constant temperature shaker for 2 h, the supernatant was taken to determine the absorbance after solid–liquid separation, and then the concentration of the remaining reagent in the solution was calculated by the standard curve.

2.5. FTIR Measurements

Put 2.00 g single mineral sample (5 μm) into a beaker, pour 30 mL deionized water into it, add flotation agent and stir for 1 h. The mineral samples were then washed three times with deionized water. The last step is to dry in a vacuum oven (40 °C). The sample was prepared by the potassium bromide (KBr) disk method, and FTIR spectra were obtained by the IRffinity-1 Fourier Transform Infrared Spectrometer (Shimadzu, Kyoto, Japan). The region of spectral data is 400–4000 cm^{-1} [26].

2.6. XPS Analysis

Samples for XPS testing were prepared in the same manner as FTIR. XPS spectra were obtained by measuring the sample with the Thermo ScientificESCALab 250Xi XPS tester. The XPS spectra of mineral samples, before and after being treated by flotation reagent, were analyzed, and the specific adsorption sites between reagent molecules and minerals were determined.

3. Results

3.1. Micro-Flotation Results

3.1.1. Single Mineral Flotation Tests

Figure 4a shows the relationship between the collector SHA dose and mineral floatability at pH 8. The results show that when the dose of collector SHA is 4×10^{-4} mol/L, bastnaesite and fluorite reach their maximum values [10]. The results in Figure 4b show that bastnaesite and fluorite have high floatability under weak alkaline conditions with/without SS. When the pH is 8~10, the floatability of bastnaesite is better, and the floatability of fluorite is better at pH 7~8. At pH 10, the presence of SS has no obvious effect on the flotation recovery of bastnaesite, but the flotation recovery of fluorite decreases greatly. This result shows that SS inhibits fluorite in this system [10,27]. Figure 4c shows the SS dose as a function of minerals floatability at pH 10. When the dose of SS is low, the floatability of bastnaesite decreases slightly, and the floatability of fluorite decreases rapidly. With increasing SS, the recovery of bastnaesite and fluorite obviously decrease. When the dose of SS is 40 mg/L, the floatability of bastnaesite and fluorite varies greatly, and selective inhibition is the most obvious.

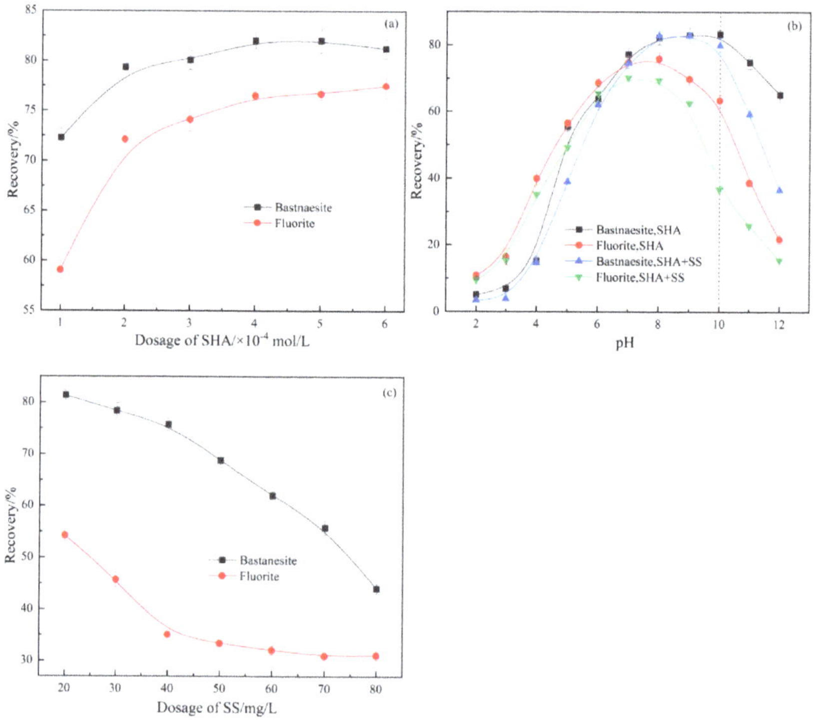

Figure 4. Recoveries of bastnaesite and fluorite as a function of dosages of SHA (**a**: Terpineol 20 mg/L; pH 8), pH (**b**: Terpineol 20 mg/L; SHA 4 × 10^{-4} mol/L; SS 40 mg/L) and dosages of SS (**c**: Terpineol 20 mg/L; SHA 4 × 10^{-4} mol/L; pH 10), respectively.

3.1.2. Flotation Separation Tests of Mixed Ore

Pure mineral flotation experiments show that the best flotation separation conditions of bastnaesite and fluorite were 4 × 10^{-4} mol/L of SHA, 40 mg/L of SS and 20 mg/L of terpineol, pH 10. In order to further determine the flotation results of a single mineral, the bastnaesite and fluorite were mixed manually, according to the mass ratio of 1:1, and the mixed mineral was used for flotation. Table 3 shows the flotation test results.

Table 3. The results of flotation separation of artificial mixed ore.

Reagents	Product	Yield (%)	Grade (%)	Recovery (%)
			Bastnaesite	Bastnaesite
pH: 10; SHA: 4 × 10^{-4} mol/L; T: 20 mg/L	Concentrate	69.23	57.28	79.34
	Tailing	30.77	33.56	20.66
	Feed	100.00	49.98	100.00
pH: 10; SHA: 4 × 10^{-4} mol/L; SS: 40 mg/L; T:20 mg/L	Concentrate	49.67	82.08	83.21
	Tailing	46.99	17.01	16.79
	Feed	100.00	49.98	100.00

Table 3 shows that bastnaesite recovery is 79.34% in the concentrate without SS, but the grade of bastnaesite is only 61.28%. After adding 40 mg/L SS, the recovery and grade of bastnaesite increased to 83.21% and 82.08%, respectively. These figures indicate that SS can effectively separate bastnaesite and fluorite at pH 10.

3.2. Potentiodynamic Measurements

The change of the Zeta potential on the mineral surface can indirectly represent the adsorption of flotation agents on the mineral surface under different pH conditions [28,29]. Figure 5 shows that the isoelectric points of bastnaesite and fluorite are 7.21 and 9.12, respectively, demonstrating the Zeta potentials of bastnaesite and fluorite before and after interaction with different reagents [28,30]. When SHA was used as the sole treatment, the Zeta potentials of bastnaesite and fluorite were negatively shifted as a result of adsorption between SHA and the mineral. After the zero point, the Zeta potentials of bastnaesite and fluorite were still negatively shifted, indicating chemisorption between SHA and the mineral [26,31].

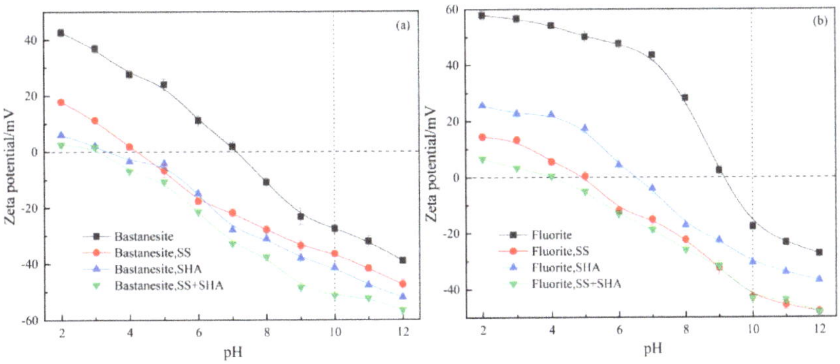

Figure 5. Zeta potentials of the minerals before (**a**) and after (**b**) treatment with reagents (SHA 4×10^{-4} mol/L; SS 20 mg/L).

When SS was added alone, the Zeta potential of bastnaesite shifted slightly in a negative direction, indicating an interaction between the SS components and bastnaesite. The negative shift of the Zeta potential of fluorite over the whole pH range indicates that the interaction between the SS and fluorite is stronger, and there is also chemisorption, in addition to electrostatic adsorption [32]. More precisely, at pH 10, the Zeta potential of the bastnaesite shifts by −6.43 mV (from −29.23 mV to −35.66 mV), and the Zeta potential of the fluorite shifts by −25.48 mV (from −18.91 mV to −44.39 mV).

When adding SS and SHA, the Zeta potential of the bastnaesite still shows a significant negative shift at pH 10 from −29.23 mV to −53.11 mV. These results show that SHA can adsorb on the surface of bastnaesite, even in the presence of SS, and the Zeta potential shift of bastnaesite is more obvious under the action of SHA, which indicates that the adsorption capacity of SS on the surface of bastnaesite is weaker than SHA, which is the reason for the good floatability of bastnaesite at a low SS concentration [33,34]. However, the Zeta potential after both SS and SHA treatment of the fluorite was almost the same as that of the fluorite interacting with SS alone, indicating that adsorption on the SS fluorite surface was stronger than that on SHA. The results show that the presence of SS hinders the further adsorption of SHA on the fluorite surface. Therefore, SS has a stronger inhibitory effect on fluorite than bastnaesite.

3.3. Adsorption Capacity and Solution Chemistry

Figure 6 shows the experimental results of SHA adsorption on the bastnaesite and fluorite surfaces with/without SS. Figure 6a shows that without SS, the maximum adsorption amount of fluorite is at pH 7–8, and the maximum adsorption amount of bastnaesite is at pH 8–10. When adding SS and pH > 9.5, the presence of SS leads to a significant decrease in the adsorption amount of SHA on the fluorite surface. The adsorption experiments shown in Figure 6b investigate the effect of SS dose on SHA adsorption at pH 10. When the dose of SS is low, SHA adsorption on the surface of the bastnaesite is slightly reduced, while SHA adsorption on the surface of the fluorite is greatly reduced. The inhibition of SHA adsorption on the fluorite surface by SS indicates that SS has a stronger adsorption capacity on fluorite than SHA, which may also explain the results of the single mineral flotation experiments. This phenomenon may be due to the chemical reaction between SS and the fluorite surface, resulting in more stable substances.

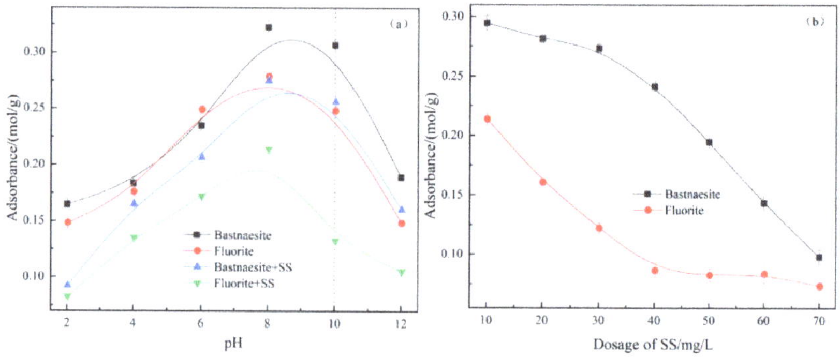

Figure 6. Adsorption of bastnaesite and fluorite as a function of pH (**a**: SHA 4×10^{-4} mol/L; SS 20 mg/L), and dosages of SS (**b**: SHA 4×10^{-4} mol/L; pH 10), respectively.

In the flotation process, the inhibitor in pulp shows various shapes under different pH conditions, and the form of the inhibitor has a certain influence on the flotation performance [29,31,35]. The solution chemistry of SS can be analyzed to obtain the distribution coefficients of each component at different pH values. Figure 7 shows the functional relationship between the hydrolysis components of SS and pH. When the pH < 9.5, the main component of SS in the solution is $Si(OH)_4$; when the pH is 9.5–12.4, the main component of SS in the solution is $SiO(OH)_3^-$ at pH 9.5–12.5; and when pH $\geq$ 12.5, the main component of SS in the solution is $SiO_2(OH)_2^{2-}$ [36,37]. When SS acts, the pH of the slurry is 9.5–12, and the main component in the solution is $SiO(OH)_3^-$. Silicate particles adsorb on the fluorite surface with high solid strength and a large adsorption capacity, which reduces fluorite recovery. In contrast, the adsorption capacity of bastnaesite is small, and the fixation strength is weak. Thus, SS has good selectivity [35,38]. At pH 9.5–12.5, SS has the strongest adsorption capacity on the fluorite surface, which explains why SHA adsorbed on the fluorite surface decreases significantly in this pH range and further verifies the results of the single mineral flotation test.

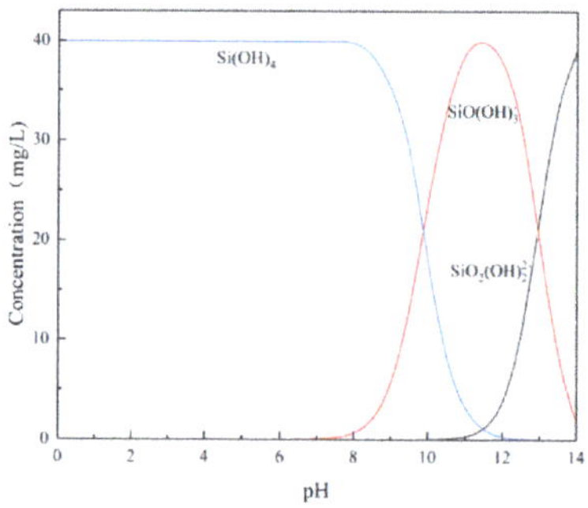

Figure 7. Distribution diagram of silicate ions as a function of pH in 40 mg/L SS.

3.4. Infrared Spectroscopy Analysis

The interference of SS adsorption on bastnaesite and fluorite on SHA was studied by infrared spectrum analysis of SS, SHA and treated bastnaesite and fluorite. The spectra of SHA and SS are shown in Figure 8. Table 4 shows the main bands corresponding to the relevant chemical bonds. In the FTIR spectra of SHA, 3283 cm^{-1} and 3050 cm^{-1} are the stretching vibrations of -OH with CH_3 on the benzene ring [38,39]; the spectra of 1521 cm^{-1} and 1153 cm^{-1} correspond to the stretching vibrations of benzene rings C=C and C=O, respectively. The spectra at 903 cm^{-1} and 1031 cm^{-1} are the asymmetrical and symmetrical tensile vibrations of -C-N and =N-O, respectively, which are major functional groups of SHA [39]. In the infrared spectrum of SS, the absorption peaks of 3335 cm^{-1} and 1660 cm^{-1} are the -OH stretching vibration and bending vibration absorption peaks of SS, and the absorption peaks of 995 cm^{-1} are the Si-O stretching vibration absorption peaks of SS at 995 cm^{-1} [40].

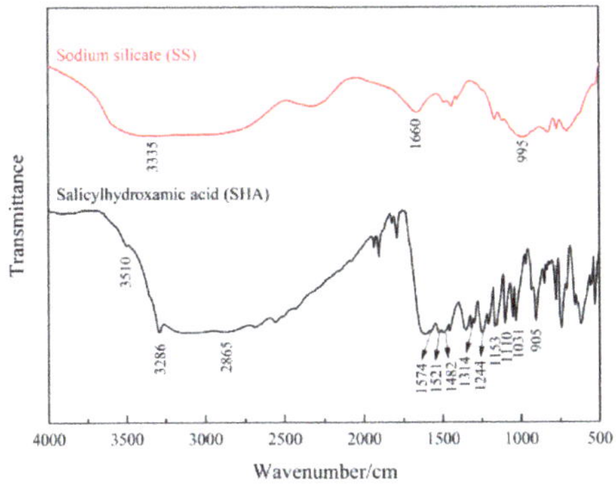

Figure 8. The infrared spectra of SHA and SS.

Table 4. The bands corresponding to the relevant chemical bond in FT-IR spectra of the minerals.

FTIR	Band (cm^{-1})	Chemical Bond
SHA	3286	O-H stretching vibration
	2865	-CH_3 stretching vibrations
	1574	C=C stretching vibrations
	1153	C-O stretching vibrations
	1031	=N-O anti-symmetric stretching vibration
	905	N-O symmetric stretching vibration
Bastnaesite	3584	-OH stretching vibration
	1824	C=O and cation coordination
	1445	CO_3^{2-} anti-symmetric stretching vibration
	1087	CO_3^{2-} symmetric stretching vibration
	867	CO_3^{2-} plane bending vibration
	723	CO_3^{2-} in-plane bending vibration
Bastnaesite + SS + SHA	3793	N-H stretching vibrations
	3696	C-H stretching vibrations
	1035	N-O stretching vibrations
	1018	N-O stretching vibrations

Figure 9a shows the infrared spectra of SHA- and SS-treated bastnaesite. The characteristic peak at 3584 cm^{-1} in the pure bastnaesite is due to the -OH stretching vibration from water in the air or on the surface of the sample, which indicates that the ions are hydroxylated at pH 10 and that the planar bending vibration and asymmetric stretching vibration of 2503 cm^{-1} in bastnaesite occur [41]. The peaks at 1824 cm^{-1} and 1763 cm^{-1} are attributed to stretching caused by the coordination between the C=O bond and rare-earth element cations [39]. The bands at 1445 cm^{-1} and 1085 cm^{-1} are attributed to the asymmetric stretching of carbonate. The bands at 867 cm^{-1} and 723 cm^{-1} are the out-of-plane and in-plane bending of carbonate [39,42]. After the ore is treated with SS alone, there is no

SS band, indicating that SS is not obviously adsorbed on the bastnaesite surface. After SS and SHA treatment, the spectrum of bastnaesite shows the characteristic peaks of tensile vibration of C-O and N-H at 3793 cm^{-1}, 3696 cm^{-1}, 3660 cm^{-1} and 3584 cm^{-1}, and the characteristic peaks of out-of-plane bending of =N-O are at 1035 cm^{-1} and 1018 cm^{-1} [39,41,42]. The results show that the presence of SS does not affect the adsorption of SHA on the bastnaesite surface.

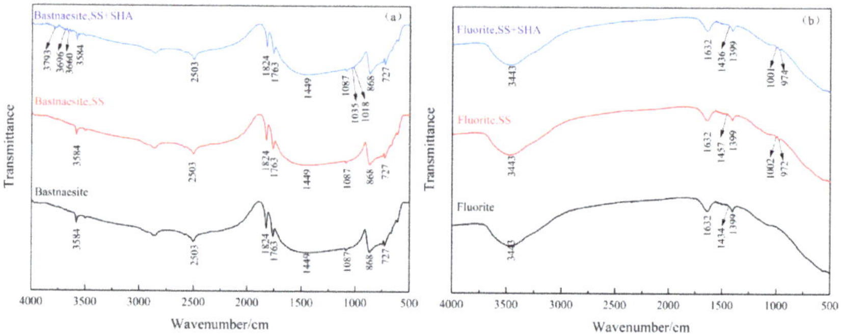

Figure 9. FTIR spectra of pure, SS- and SS+ SHA-treated bastnaesite (**a**) and pure, SS- and SS+ SHA-treated fluorite (**b**).

Figure 9b shows the infrared spectrum of fluorite before and after SS and SHA treatment. For fluorite, the broad peaks near 3443 cm^{-1} and 1632 cm^{-1} are attributed to adsorbed water molecules, and the characteristic peaks at 1434 cm^{-1} and 1399 cm^{-1} are attributed to fluorite [43,44]. After SS treatment, there are new characteristic peaks at 1002 cm^{-1} and 974 cm^{-1} on the fluorite surface, which may be caused by the tensile vibration of Si-O, indicating that there are $SiO(OH)_3^-$ groups on the fluorite surface [45]. Moreover, the characteristic peak of fluorite at 1434 cm^{-1} moves to 1457 cm^{-1}, indicating that SS is adsorbed on the fluorite surface [45,46]. There are no new characteristic peaks on the fluorite surface treated with SHA and SS, unlike that of the surface treated with SS alone. FTIR analysis showed that the surface of the bastnaesite treated with SS did not affect the further adsorption of SHA. When SS and SHA are both used as treatments, the IR spectra of the fluorite are very similar to those of the SS alone. SS can hinder SHA adsorption behavior on the fluorite surface, which further confirms that SS selectively inhibits the fluorite surface, thereby reducing fluorite floatability.

3.5. XPS Measurement Results

Table 5 shows the change of the relative atomic concentration before and after the interaction of the flotation reagent with bastnaesite and fluorite. After SS treatment, the atomic concentrations of C, Ce, La and F decreased slightly, while the atomic concentrations of O and Si only slightly increase [47,48]. After SS and SHA treatment, the C and N atomic concentrations increase by 7.43% and 2.12%, respectively; the Si atomic concentration increases slightly; the O, Ce and La atomic concentrations decrease by 7.28%, 1.00% and 1.36%, respectively; and the F atomic concentration decreases slightly. When compared with bastnaesite, the relative atomic concentration of the fluorite surface treated with SS changes significantly. The atomic concentrations of F, Ca and C decrease by 3.59%, 1.56% and 0.7%, respectively, and the atomic concentrations of O and Si increase by 3.83% and 2.02%, respectively. Especially, significant changes in the Si content suggest that a lot of SS is adsorbed on the fluorite surface. Consequently, the adsorption behavior of the flotation reagents on the mineral surface was further studied. Figure 10 shows the XPS spectra of minerals treated with/without flotation reagents. Figure 10a shows that after adding SS alone, there is no obvious change in the spectrum of the bastnaesite. After the interaction between SS and SHA, a new N 1s peak appears on the surface of the bastnaesite, and SS does not affect the adsorption of SHA on the surface of the bastnaesite. Figure 10b shows that after adding SS alone, the spectrum of the fluorite shows a new peak at Si 2p, and there are no new peaks where SS and SHA coexist, indicating that a large amount of SS covers the surface of the fluorite, and the presence of SS hinders the adsorption of SHA on the fluorite.

Table 5. Atomic concentrations of various elements on the surfaces of bastnaesite and fluorite untreated and treated by flotation reagents.

Samples	Atomic Concentration/%							
	C	N	O	Ce	La	Ca	F	Si
Bastnaesite	32.51	-	44.70	4.53	6.65	-	11.61	-
SS treated bastnaesite	31.96	-	45.12	4.49	6.61	-	10.84	0.98
SS + SHA treated bastnaesite	39.94	2.12	37.42	3.53	5.29	-	11.01	0.69
Fluorite	13.57	-	5.16	-	-	27.63	53.64	-
SS treated fluorite	12.87	-	8.99	-	-	26.07	50.05	2.02
SS + SHA treated fluorite	12.93	0.12	8.97	-	-	26.02	49.93	2.03

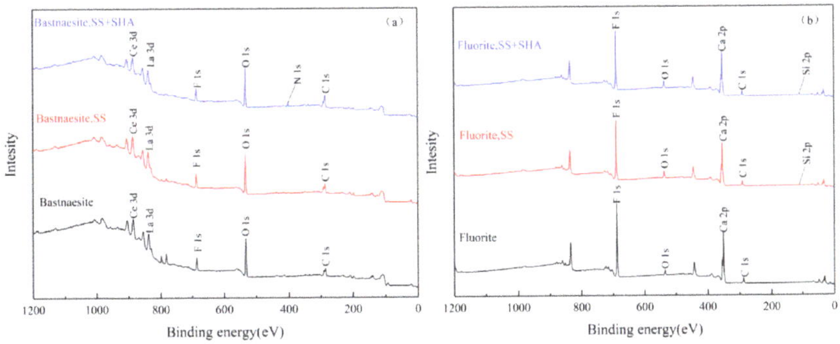

Figure 10. XPS spectra of pure, SS- and SS+ SHA-treated bastnaesite (**a**) and pure, SS- and SS+ SHA-treated fluorite (**b**).

Figure 11a shows the high-definition Ce $3d_{5/2}$ XPS spectra of the pure and flotation reagent-treated bastnaesite samples, and the detailed analysis results of the Ce $3d_{5/2}$ XPS spectra are shown in Table 6. In the pure bastnaesite spectra, the three-dimensional Ce spectrum consists of spin-orbital split $3d_{5/2}$ and $3d_{3/2}$ nuclear pores, and $3d_{5/2}$ is closely linked to the $3d_{3/2}$ spectrum. In general, the binding energy of $3d_{3/2}$ is 18–19 eV higher than that of $3d_{5/2}$, and the intensity ratio between $3d_{5/2}$ and $3d_{3/2}$ is 1.5 [49,50]. For the sake of clarity, only the Ce $3d_{5/2}$ spectrum is analyzed and displayed in this study. The peaks of 888.18 eV and 888.18eV are more likely to be related to Ce (III), which may be due to the complex electron configuration of Ce atoms and the effect of F atoms on the crystal structure of bastnaesite [50,51]. The Ce $3d_{5/2}$ slightly shifted (<0.10) after adding SS alone, all within the instrument error range [52]. When SS and SHA were added together, the Ce $3d_{5/2}$ peak near 531.55 eV decreased by 0.16 ± 0.02 eV. The decrease in binding energy indicates that the SHA is chemically adsorbed on the Ce ions on the surface of the bastnaesite.

Figure 11b shows the high-resolution C 1s XPS spectra of the pure and flotation reagent-treated bastnaesite samples, and the detailed analysis results of the C 1s XPS spectra are shown in Table 7. The results show that the C 1s XPS spectra of pure bastnaesite have two peaks at the binding energies of C-C and CO_3^{2-} of 284.77 eV and 289.31 eV, respectively. The results show that the C 1s XPS spectra of pure bastnaesite can have two component peaks at binding energies of 284.77 eV and 289.31 eV from C-C and CO_3^{2-}, respectively [53,54]. In the C 1s XPS spectra of the bastnaesite treated by SS alone, there is no obvious shift of the C 1s peak, while the peaks at 284.77 eV and 289.31 eV move forward by 0.24 eV and 0.40 eV, respectively, after SS and SHA treatment. This shift may be due to the superposition of the SHA C=O group with CO_3^{2-} on the surface of the bastnaesite [55].

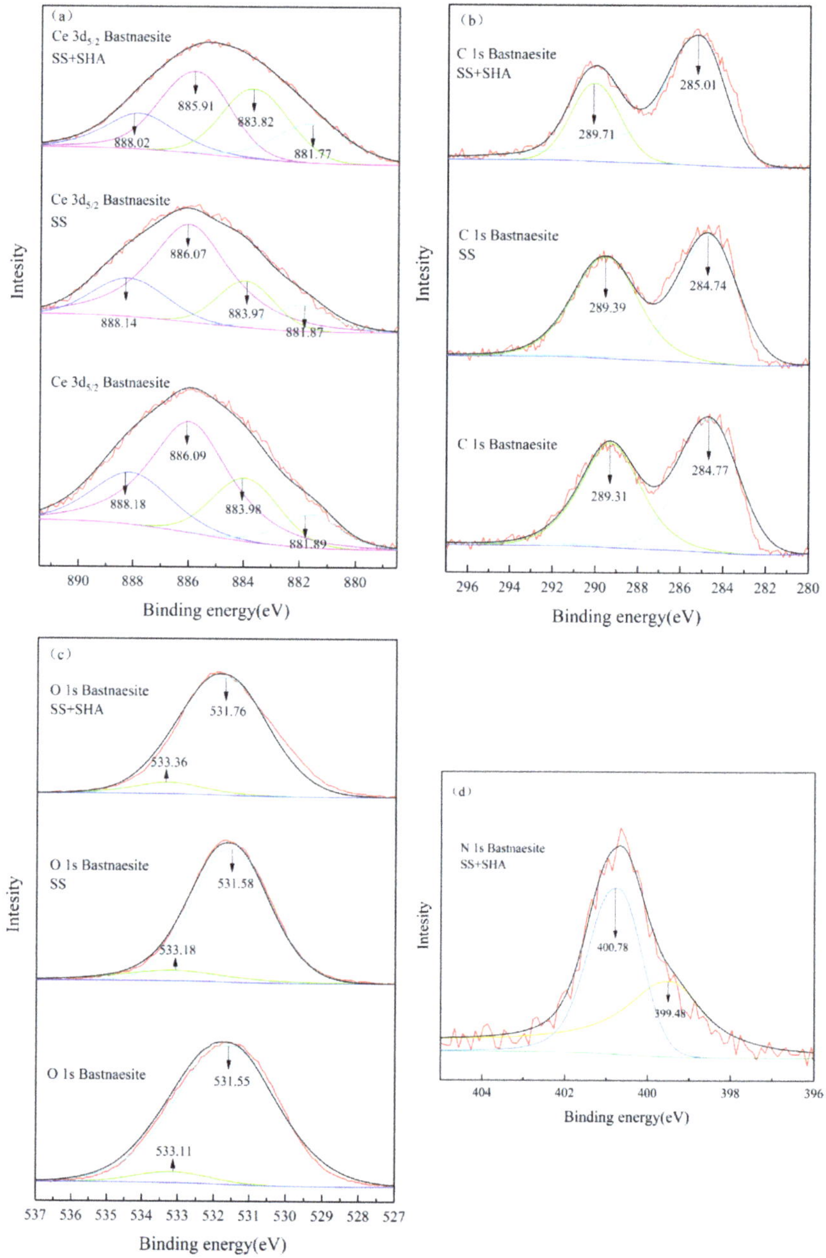

Figure 11. High-resolution XPS spectra of Ce $3d_{5/2}$ (**a**), C 1s (**b**), O 1s (**c**) and N 1s (**d**) of pure, SS- and SS+ SHA-treated bastnaesite particles, respectively.

Table 6. Analysis results of Ce $3d_{5/2}$ XPS spectra of bastnaesite samples untreated and treated by flotation reagent.

Samples	Binding Energy/eV	FWHM/eV	Chemical Shift/eV
Bastnaesite	881.89	2.54	/
	883.98	2.54	/
	886.09	2.54	/
	888.18	2.54	/
Bastnaesite + SS	881.87	2.51	−0.02
	883.97	2.52	−0.01
	886.07	2.51	−0.02
	888.14	2.53	−0.04
Bastnaesite + SS + SHA	881.77	2.63	−0.12
	883.82	2.61	−0.16
	885.91	2.63	−0.18
	888.02	2.63	−0.16

Table 7. Analysis results of C 1s XPS spectra of bastnaesite samples untreated and treated by flotation reagent.

Samples	Binding Energy/eV	FWHM/eV	Chemical Shift/eV
Bastnaesite	284.77	1.81	/
	289.31	1.83	/
Bastnaesite + SS	284.74	1.81	−0.03
	289.39	1.80	0.08
Bastnaesite + SS + SHA	285.01	1.86	0.24
	289.71	1.87	0.40

Figure 11c shows the high-definition O 1s XPS spectra of the pure and flotation reagent-treated bastnaesite samples, and the detailed analysis results of the O 1s XPS spectra are shown in Table 8. The results show that the O 1s region of the bastnaesite can be fitted to two central peaks of binding energy of 531.55 eV and 533.11 eV in the XPS spectra, respectively, from CO_3^{2-} and Ce-OH [55,56]. In the O 1s XPS spectrum of the bastnaesite treated by SS alone, the O 1s XPS peak of CO_3^{2-} moves 0.10 in the positive direction. In the O 1s XPS spectrum of SS- and SHA-treated bastnaesite, the O 1s XPS peak of CO_3^{2-} shifted to the positive direction by 0.24 eV, which is due to the benzene-ring-binding -OH group in SHA [55]. The O 1s XPS peak in the Ce-OH shifts 0.20 eV in the positive direction due to the reaction of the C=O group of the SHA with the surface of the bastnaesite [57].

Table 8. Analysis results of O 1s XPS spectra of bastnaesite samples untreated and treated by flotation reagent.

Samples	Binding Energy/eV	FWHM/eV	Chemical Shift/eV
Bastnaesite	531.55	1.73	/
	533.11	1.63	/
Bastnaesite + SS	531.58	1.48	0.03
	533.18	1.34	0.07
Bastnaesite + SS + SHA	531.76	1.78	0.21
	533.36	1.67	0.25

In addition, Figure 11d shows the high-definition N 1s XPS spectra of the flotation reagent-treated bastnaesite samples, and the detailed analysis results of N 1s XPS spectra are shown in Table 9. It is further confirmed that NHA is adsorbed on the bastnaesite surface. This adsorption may be attributed to the N atom in the =N-OH forming a new saturated chelate with the surface of the bastnaesite [24,58].

Table 9. Analysis results of N 1s XPS spectra of bastnaesite samples untreated and treated by flotation reagent.

Samples	Binding Energy/eV	FWHM/eV	Chemical Shift/eV
Bastnaesite + SS + SHA	399.48 400.78	1.57 1.47	399.48 400.78

Figure 12a shows the high-definition Ca 2p XPS spectra of the pure and flotation reagent-treated fluorite samples, and the detailed analysis results of the N 1s XPS spectra are shown in Table 10. In the spectra of the pure fluorite samples, the binding energies at 347.74 eV and 351.25 eV are the peaks of Ca $2p_{3/2}$ and Ca $2p_{1/2}$, respectively. In the XPS spectra of the fluorite samples treated by SS alone, the binding energy offsets of the Ca $2p_{3/2}$ and Ca $2p_{1/2}$ peaks are 0.39 eV and 0.27 eV (>0.20 eV), respectively. The obvious shift of the Ca 2p energy spectrum on the fluorite surface indicates that the interaction between the fluorite and the SS is chemical bonding rather than simple physical bonding. This observation shows the chemical surroundings of the fluorite surface has changed. However, the Ca 2p peak of the SS- and SHA-treated fluorite shows no significant change compared with the SS alone [59].

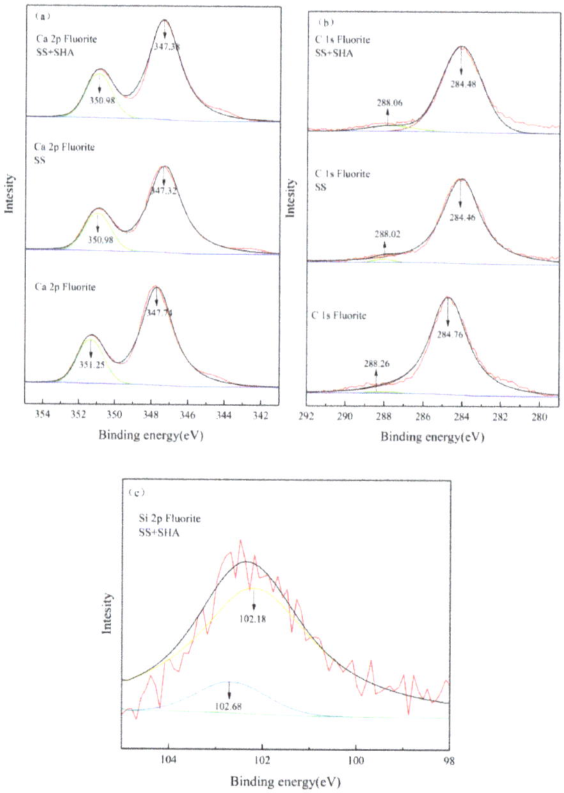

Figure 12. High-resolution XPS spectra of Ca 2p (**a**), C 1s (**b**) and Si 2p (**c**) of pure, SS- and SS+SHA-treated fluorite particles, respectively.

Table 10. Analysis results of Ca 2p XPS spectra of fluorite samples untreated and treated by flotation reagent.

Samples	Binding Energy/eV	FWHM/eV	Chemical Shift/eV
Fluorite	347.74 351.25	2.05 2.05	/ /
Fluorite + SS	347.32 350.98	2.24 2.24	−0.42 −0.27
Fluorite + SS + SHA	347.38 350.98	2.15 2.15	−0.36 −0.27

Figure 12b shows the high-definition C 1s XPS spectra of the pure and flotation reagent-treated fluorite samples, and the detailed analysis results of the C 1s XPS spectra are shown in Table 11. In the spectra of the pure fluorite samples, the C 1s peak is attributed to CO_3^{2-} [24]. In the XPS spectra of the fluorite samples treated with SS alone, the binding energy of the C 1s peak shifts to 0.30 eV and 0.24 eV [24,59]. The fluorite samples treated with SS and SHA did not show the characteristic peaks of the carbonyl groups and the benzene ring of SHA. The binding energy offset of the C 1s peak is less than 0.10 eV, which is within the range of instrument error.

Figure 12c shows the high-definition Si 2p XPS spectra of the flotation reagent-treated fluorite samples, and the detailed analysis results of the Si 2p XPS spectra are shown in Table 12. At 102.16 ev and 102.69 ev, the Si 2p peak of the fluorite was fitted by double peaks [35,37,45]. Because the binding energy of the Si-C bond does not exceed 102 eV near 100.5 eV, it is possible that the SS interacts with the Ca^{2+} on the surface of the fluorite to form a new chemical bond [37,45,60], indicating that the SS is adsorbed on the fluorite. The results show that the SS has chemisorption on the fluorite surface, and the adsorption capacity is higher than that of the bastnaesite.

Table 11. Analysis results of C 1s XPS spectra of fluorite samples untreated and treated by flotation reagent.

Samples	Binding Energy/eV	FWHM/eV	Chemical Shift/eV
Fluorite	284.76 288.26	2.21 2.21	/ /
Fluorite + SS	284.46 288.02	2.44 2.44	−0.30 −0.24
Fluorite + SS + SHA	284.48 288.06	3.88 3.88	−0.28 −0.20

Table 12. Analysis results of Si 2p XPS spectra of fluorite samples untreated and treated by flotation reagent.

Samples	Binding Energy/eV	FWHM/eV	Chemical Shift/eV
Fluorite + SS + SHA	102.18 102.68	2.01 1.51	102.18 102.68

3.6. Discussion

By previous detection and analysis, we found that SS chelates strongly with Ca^{2+} on the fluorite surface at pH 10, but has little interaction with Ce^{3+} on the bastnaesite surface. The chelation reaction between SS and calcium ions was formulated as follows. In each cell, $SiO(OH)_3^-$ provides a pair of oxygen atoms with Ca^{2+} to form a chelation.

$$Ca^{2+}{}_{<surf.>} + SiO(OH)_3^- \longrightarrow Ca^{2+} - OSi(OH)_{3<surf.>} \quad (2)$$

$$Ca^{2+} - F_{<surf.>} + NaSiO(OH)_3^- \longrightarrow Ca^{2+} - OSi(OH)_{3<surf.>} \quad (3)$$

Under the alkaline condition (pH > 10), the $SiO(OH)_3^-$ concentration in the solution is higher, SS can be firmly adsorbed on the fluorite surface, and then the fluorite flotation has a fine inhibition effect [19,37,38,45]. These results indicate that SS is more easily adsorbed on the fluorite surface under specific pH conditions, and a large amount of SS on the fluorite surface interferes with further adsorption of SHA on the fluorite surface [37]. Conversely, only a little SS was adsorbed on the surface of bastnasite, which did not significantly affect the further adsorption of SHA on the surface of the bastnasite. The adsorption model of flotation reagents on the surface of two minerals is shown in Figure 13.

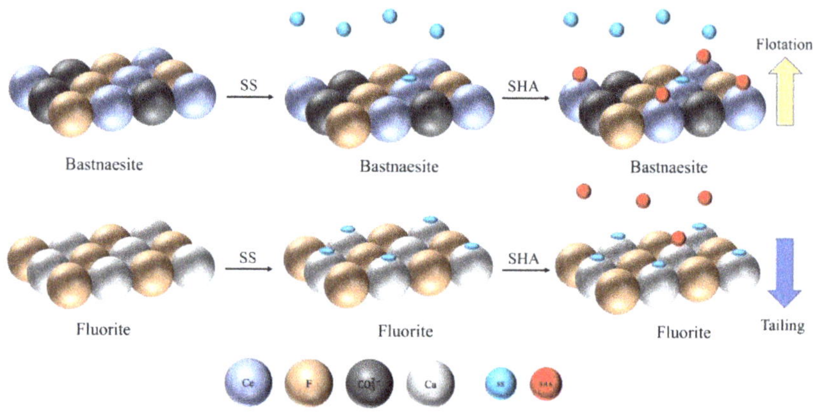

Figure 13. Reaction of flotation reagents with the bastnaesite and fluorite surfaces.

4. Conclusions

The efficient and economical inhibitor SS has been shown to promote the separation of bastnaesite from fluorite. The following is a summary of the study:

(1) The results of micro-flotation show that SS has strong selective inhibition of fluorite in a SHA flotation system. At a pH of 10, SS dosage 40 mg/L and SHA dosage 4×10^{-4} mol/L, the flotation separation effect of bastnaesite and fluorite can be improved, and accurate recovery of bastnaesite is realized.

(2) The potentiodynamic potential and adsorption capacity showed that at pH = 10, the presence of SS did not affect the further adsorption of SHA on the surface of bastnaesite, but the adsorption of SHA on the surface of fluorite was greatly hindered, which may be caused by the selective adsorption of $SiO(OH)_3^-$ ions on the surface of fluorite. Under different pH conditions, the adsorption behavior of hydrolysates of SS on the surface of the two minerals with different strengths leads to the difference in the final floatability of the two minerals, which is the key to achieving flotation separation.

(3) FTIR and XPS analysis showed that SS has a strong binding force with Ca^{2+} on the fluorite surface, while Ce^{3+} had a weak reaction with the bastnaesite surface. This also explains that the selective inhibition of SS on fluorite in flotation separation is determined by the different intensity of action of $SiO(OH)_3^-$ ions on the active sites on the surface of the two minerals. This study has theoretically verified the reason why SS can selectively inhibit bastnaesite and fluorite in flotation separation, and provides guidance for the design and development of flotation separation processes in the future.

Author Contributions: M.W.: Data curation, Investigation, Visualization, Writing original draft. W.X.: Writing—review & editing. J.X.: Methodology, Investigation. Y.G. and J.D.: Software, Validation. D.C.: Funding acquisition, Supervision. A.O. and M.L.: Visualization. L.Z.: Funding acquisition, Supervision. M.L. All authors have read and agreed to the published version of the manuscript.

Funding: The authors gratefully acknowledge the financial support of the Youth Talent Project of the China Ministry of Natural Resources, grant number B1901, 12110600000018003914.

Acknowledgments: The authors gratefully acknowledge the financial support of the Youth Talent Project of the China Ministry of Natural Resources, grant number B1901, 12110600000018003914.

Conflicts of Interest: The authors declare that they have no known competing financial interests or personal relationships that could have appeared to influence the work reported in this paper.

References

1. Zhang, W.; Rezaee, M.; Bhagavatula, A.; Li, Y.; Groppo, J.; Honaker, R. A review of the occurrence and promising recovery methods of rare earth elements from coal and coal by-products. *Int. J. Coal Prep. Util.* **2015**, *35*, 295–330. [CrossRef]
2. Filippov, L.O.; Dehaine, Q.; Filippova, I.V. Rare earths (La, Ce, Nd) and rare metals (Sn, Nb, W) as by-products of kaolin production—Part 3: Processing of fines using gravity and flotation. *Miner. Eng.* **2016**, *95*, 96–106. [CrossRef]
3. Yu, B.; Aghamirian, M. REO mineral separation from silicates and carbonate gangue minerals. *Can. Metall. Q.* **2015**, *54*, 377–387. [CrossRef]
4. Yu, B.; Hu, Z.; Zhou, F.; Feng, J.; Chi, R. Lanthanum (III) and Yttrium (III) Adsorption on Montmorillonite: The Role of Aluminum Ion in Solution and Minerals. *Miner. Process. Extr. Metall. Rev.* **2020**, *41*, 107–116. [CrossRef]
5. Archambo, M.S.; Kawatra, S.K. Extraction of Rare Earths from Red Mud Iron Nugget Slags with Oxalic Acid Precipitation. *Miner. Process. Extr. Metall. Rev.* **2022**, *43*, 656–663. [CrossRef]
6. Talan, D.; Huang, Q.; Liang, L.; Song, X. Conceptual Process Development for the Separation of Thorium, Uranium, and Rare Earths from Coarse Coal Refuse. *Miner. Process. Extr. Metall. Rev.* **2022**, *149*, 1–16. [CrossRef]
7. Cao, S.; Cao, Y.; Ma, Z.; Liao, Y. Metal ion release in bastnaesite flotation system and implications for flotation. *Minerals* **2018**, *8*, 203. [CrossRef]
8. Xiao, J.; Zou, K.; Zhong, N.; Gao, D. Selective separation of iron and scandium from Bayer Sc-bearing red mud. *J. Rare Earths* 2022, *in press*. [CrossRef]
9. Dutta, T.; Kim, K.H.; Uchimiya, M.; Kwon, E.E.; Jeon, B.H.; Deep, A.; Yun, S.-T. Global demand for rare earth resources and strategies for green mining. *Environ. Res.* **2016**, *150*, 182–190. [CrossRef]
10. Jordens, A.; Cheng, Y.P.; Waters, K.E. A review of the beneficiation of rare earth element bearing minerals. *Miner. Eng.* **2013**, *41*, 97–114. [CrossRef]
11. Nduwa-Mushidi, J.; Anderson, C.G. Surface Chemistry and Flotation Behaviors of Monazite–Apatite–Ilmenite–Quartz–Rutile–Zircon with Octanohydroxamic Acid. *J. Sustain. Metall.* **2017**, *3*, 62–72. [CrossRef]
12. Karshigina, Z.; Abisheva, Z.; Bochevskaya, Y.; Akcil, A.; Sargelova, E.; Sukurov, B.; Silachyov, I. Recovery of rare earth metals (REMs) from primary raw material: Sulphatization-leaching-precipitation-extraction. *Miner. Process. Extr. Metall. Rev.* **2018**, *39*, 319–338. [CrossRef]
13. Jordens, A.; Marion, C.; Kuzmina, O.; Waters, K.E. Surface chemistry considerations in the flotation of bastnäsite. *Miner. Eng.* **2014**, *66*, 119–129. [CrossRef]
14. Kupka, N.; Rudolph, M. Froth flotation of scheelite—A review. *Int. J. Min. Sci. Technol.* **2018**, *28*, 373–384. [CrossRef]
15. Fuerstenau, D. The Adsorption of Hydroxamate on Semi-Soluble Minerals. Part I: Adsorption on Barite, Calcite and Bastnaesite. *Colloids Surf.* **1893**, *8*, 103–119.
16. Wang, Z.; Wu, H.; Yang, J.; Tang, Z.; Luo, L.; Shu, K.; Xu, Y.; Xu, L. Selective flotation separation of bastnaesite from calcite using xanthan gum as a depressant. *Appl. Surf. Sci.* **2020**, *512*, 145714. [CrossRef]
17. Xiong, W.; Wang, M.; Xiao, J.; Chen, D. Selective Adsorption of 2-Hydroxy-3-Naphthalene Hydroxamic Acid on the Surface of Bastnaesite and Calcite. *Minerals* **2022**, *12*, 1341. [CrossRef]
18. Yang, Z.; Bian, X.; Wu, W. Flotation performance and adsorption mechanism of styrene phosphonic acid as a collector to synthetic $(Ce,La)_2O_3$. *J. Rare Earths* **2017**, *35*, 621–628. [CrossRef]
19. Marinakis, K.I.; Shergold, H.L. Influence of Sodium Silicate Addition on The Adsorption of Oleic Acid by Fluorite, Calcite and Barite. *Int. J. Miner. Process.* **1985**, *14*, 177–193. [CrossRef]
20. Archambo, M.; Kawatra, S.K. Red Mud: Fundamentals and New Avenues for Utilization. *Miner. Process. Extr. Metall. Rev.* **2021**, *42*, 427–450. [CrossRef]
21. Xiao, J.; Zou, K.; Gao, D.; Zhong, N.; Xiong, W.; Li, C.; Huang, W.; Liang, G. Effective Extraction of Titanium and Iron from Coarse Anatase Concentrate. *JOM* **2022**, *74*, 3833–3842. [CrossRef]
22. Li, W.; Shi, D.; Han, Y. A selective flotation of fluorite from dolomite using caustic cassava starch and its adsorption mechanism: An experimental and DFT Study. *Colloids Surf. A Physicochem. Eng. Asp.* **2022**, *633*, 127876. [CrossRef]

23. Foucaud, Y.; Badawi, M.; Filippov, L.O.; Barres, O.; Filippova, I.V.; Lebègue, S. Synergistic adsorptions of Na_2CO_3 and Na_2SiO_3 on calcium minerals revealed by spectroscopic and: Ab initio molecular dynamics studies. *Chem. Sci.* **2019**, *10*, 9928–9940. [CrossRef] [PubMed]
24. Guo, Z.; Khoso, S.A.; Wang, J.; Zhang, C.; Gao, Z.; Sun, W.; Tian, M.; Liu, Y. Interaction mechanism of 2-hydroxy-3-naphthyl hydroxamic acid and 1-hydroxy-2-naphthyl hydroxamic acid in the flotation separation of bastnaesite/fluorite: Experiments and first-principles calculations. *Sep. Purif. Technol.* **2022**, *285*, 120307. [CrossRef]
25. Qin, W.; Hu, J.; Zhu, H.; Jiao, F.; Pan, Z.; Jia, W.; Han, J.; Chen, C. Selective inhibition mechanism of PBTCA on flotation separation of magnesite from calcite. *Colloids Surf. A Physicochem. Eng. Asp.* **2021**, *630*, 127597. [CrossRef]
26. Yu, X.; Zhang, R.; Zeng, Y.; Cheng, C.; Huang, Z.; Wang, J.; He, G.; Wang, H. The effect and mechanism of cinnamic hydroxamic acid as a collector in flotation separation of malachite and calcite. *Miner. Eng.* **2021**, *164*, 106847. [CrossRef]
27. Wanhala, A.K.; Doughty, B.; Bryantsev, V.S.; Wu, L.; Mahurin, S.M.; Jansone-Popova, S.; Cheshire, M.C.; Navrotsky, A.; Stack, A. Adsorption mechanism of alkyl hydroxamic acid onto bastnäsite: Fundamental steps toward rational collector design for rare earth elements. *J. Colloid Interface Sci.* **2019**, *553*, 210–219. [CrossRef]
28. Gao, Z.; Bai, D.; Sun, W.; Cao, X.; Hu, Y. Selective flotation of scheelite from calcite and fluorite using a collector mixture. *Miner. Eng.* **2015**, *72*, 23–26. [CrossRef]
29. Devi, N.; Sukla, L.B. Studies on Liquid-Liquid Extraction of Yttrium and Separation from Other Rare Earth Elements Using Bifunctional Ionic Liquids. *Miner. Process. Extr. Metall. Rev.* **2019**, *40*, 46–55. [CrossRef]
30. Owens, C.L.; Nash, G.R.; Hadler, K.; Fitzpatrick, R.S.; Anderson, C.G.; Wall, F. Zeta potentials of the rare earth element fluorcarbonate minerals focusing on bastnäsite and parisite. *Adv. Colloid Interface Sci.* **2018**, *256*, 152–162. [CrossRef]
31. Wang, Z.; Wu, H.; Xu, Y.; Shu, K.; Fang, S.; Xu, L. The effect of dissolved calcite species on the flotation of bastnaesite using sodium oleate. *Miner. Eng.* **2020**, *145*, 106095. [CrossRef]
32. Meng, Q.; Feng, Q.; Shi, Q.; Ou, L. Studies on interaction mechanism of fine wolframite with octyl hydroxamic acid. *Miner. Eng.* **2015**, *79*, 133–138.
33. Jiang, W.; Gao, Z.; Khoso, S.A.; Gao, J.; Sun, W.; Pu, W.; Hu, Y. Selective adsorption of benzhydroxamic acid on fluorite rendering selective separation of fluorite/calcite. *Appl. Surf. Sci.* **2018**, *435*, 752–758. [CrossRef]
34. Wang, W.; Wang, H.; Wu, Q.; Zheng, Y.; Cui, Y.; Yan, W.; Deng, J.; Peng, T. Comparative study on adsorption and depressant effects of carboxymethyl cellulose and sodium silicate in flotation. *J. Mol. Liq.* **2018**, *268*, 140–148. [CrossRef]
35. Jin, S.; Ou, L.; Ma, X.; Zhou, H.; Zhang, Z. Activation mechanisms of sodium silicate-inhibited fluorite in flotation under neutral and slightly alkaline conditions. *Miner. Eng.* **2021**, *161*, 106738. [CrossRef]
36. Foucaud, Y.; Filippova, I.V.; Filippov, L.O. Investigation of the depressants involved in the selective flotation of scheelite from apatite, fluorite, and calcium silicates: Focus on the sodium silicate/sodium carbonate system. *Powder Technol.* **2019**, *352*, 501–512. [CrossRef]
37. Zhang, S.; Kuang, J.; Yu, M.; Yuan, W.; Huang, Z. Effect of ultrasonication of sodium silicate on selective adsorption of scheelite and fluorite surfaces. *Colloids Surf. A Physicochem. Eng. Asp.* **2022**, *642*, 128633. [CrossRef]
38. Natarajan, R.; Nirdosh, I. Application of topochemical, topostructural, physicochemical and geometrical parameters to model the flotation efficiencies of N-arylhydroxamic acids. *Int. J. Miner. Process.* **2003**, *71*, 113–129. [CrossRef]
39. Xiong, W.; Deng, J.; Zhao, K.; Wang, W.; Wang, Y.; Wei, D. Bastnaesite, barite, and calcite flotation behaviors with salicylhydroxamic acid as the collector. *Minerals* **2020**, *10*, 282. [CrossRef]
40. Feng, B.; Luo, X.P. The solution chemistry of carbonate and implications for pyrite flotation. *Miner. Eng.* **2013**, *53*, 181–183. [CrossRef]
41. Xiao, J.; Di, N.; Liu, G.; Zhong, H. The interaction of N-butoxypropyl-N′-ethoxycarbonylthiourea with sulfide minerals: Scanning electrochemical microscopy, diffuse reflectance infrared Fourier transform spectroscopy, and thermodynamics. *Colloids Surf. A Phys. Eng. Asp.* **2014**, *456*, 203–210. [CrossRef]
42. Xu, Y.; Xu, L.; Wu, H.; Wang, Z.; Shu, K.; Fang, S.; Zhang, Z. Flotation and co–adsorption of mixed collectors octanohydroxamic acid/sodium oleate on bastnaesite. *J. Alloys Compd.* **2020**, *819*, 152948. [CrossRef]
43. Singh, R.K. FTIR Spectroscpy of Natural Fluorite from Ambadongar, Gujarat. *J. Geol. Soc. India* **2013**, *81*, 215–218. [CrossRef]
44. Dong, L.; Jiao, F.; Qin, W.; Wei, Q. New insights into the depressive mechanism of citric acid in the selective flotation of scheelite from fluorite. *Miner. Eng.* **2021**, *171*, 107117. [CrossRef]
45. Zhang, C.; Hu, Y.; Sun, W.; Zhai, J.; Yin, Z.; Guan, Q. Effect of phytic acid on the surface properties of scheelite and fluorite for their selective flotation. *Colloids Surf. A Phys. Eng. Asp.* **2019**, *573*, 80–87. [CrossRef]
46. Foucaud, Y.; Lainé, J.; Filippov, L.; Barrès, O.; Kim, J.; Filippova, I.; Pastore, M.; Lebègue, S.; Badawi, M. Adsorption mechanisms of fatty acids on fluorite unraveled by infrared spectroscopy and first-principles calculations. *J. Colloid Interface Sci.* **2020**, *583*, 692–703. [CrossRef] [PubMed]
47. Dong, L.; Jiao, F.; Qin, W.; Zhu, H.; Jia, W. Selective depressive effect of sodium fluorosilicate on calcite during scheelite flotation. *Miner. Eng.* **2019**, *131*, 262–271. [CrossRef]
48. Nambaje, C.; Mweene, L.; Subramanian, S.; Sajeev, K.; Santosh, M. Xanthan gum based investigations into the surface chemistry of cassiterite and beneficiation of cassiterite tailings. *Miner. Process. Extr. Metall. Rev.* **2022**, *43*, 150–164. [CrossRef]
49. Qi, J.; Liu, G.; Dong, Y. Probing the hydrophobic mechanism of N-[(3-hydroxyamino)-propoxy]-N-octyl dithiocarbamate toward bastnaesite flotation by in situ AFM, FTIR and XPS. *J. Colloid Interface Sci.* **2020**, *572*, 179–189. [CrossRef]

50. Meng, D.; Wu, X.; Han, Y.; Meng, X. Polytypism and microstructures of the mixed-layer member B_2S, $CaCe_3(CO_3)_4F_3$ in the bastnaesite-(Ce)^synchysite-(Ce) series. *Earth Planet. Sci. Lett.* **2002**, *203*, 817–828. [CrossRef]
51. Qi, J.; Fan, H.; Liu, G. β-Amino-hydroxamate surfactants: Preparation, and adsorption mechanism in bastnaesite flotation. *Sep. Purif. Technol.* **2020**, *240*, 116634. [CrossRef]
52. Vickerman, J.; Gilmore, I.S. *Surface Analysis the Principal Techniques*, 2nd ed.; John Wiley & Sons: Hoboken, NJ, USA, 2011.
53. Espiritu, E.R.L.; da Silva, G.R.; Azizi, D.; Larachi, F.; Waters, K.E. The effect of dissolved mineral species on bastnäsite, monazite and dolomite flotation using benzohydroxamate collector. *Colloids Surf. A Phys. Eng. Asp.* **2018**, *539*, 319–334. [CrossRef]
54. Li, M.; Gao, K.; Zhang, D.; Duan, H.; Ma, L.; Huang, L. The influence of temperature on rare earth flotation with naphthyl hydroxamic acid. *J. Rare Earths* **2018**, *36*, 99–107. [CrossRef]
55. Światowska, J.; Lair, V.; Pereira-Nabais, C.; Cote, G.; Marcus, P.; Chagnes, A. XPS, XRD and SEM characterization of a thin ceria layer deposited onto graphite electrode for application in lithium-ion batteries. *Appl. Surf. Sci.* **2011**, *257*, 9110–9119. [CrossRef]
56. Dauscher, A.; Hilaire, L.; le Normand, F.; Müller, W.; Maire, G.; Vasquez, A. Characterization by XPS and XAS of supported Pt/TiO_2—CeO_2 catalysts. *Surf. Interface Anal.* **1990**, *16*, 341–346. [CrossRef]
57. Mullins, D.R. The surface chemistry of cerium oxide. *Surf. Sci. Rep.* **2015**, *70*, 42–85. [CrossRef]
58. Liu, S.; Liu, G.; Zhong, H.; Yang, X. The role of HABTC's hydroxamate and dithiocarbamate groups in chalcopyrite flotation. *J. Ind. Eng. Chem.* **2017**, *52*, 359–368. [CrossRef]
59. Zhang, Z.; Cao, Y.; Sun, L.; Ma, Z.; Liao, Y. Interaction forces between scheelite and fluorite in calcium solution measured by atomic force microscopy. *Appl. Surf. Sci.* **2019**, *486*, 323–336. [CrossRef]
60. Cao, Z.; Cheng, Z.; Wang, J.; Cao, Y. Synergistic depression mechanism of Ca^{2+} ions and sodium silicate on bastnaesite flotation. *J. Rare Earths* **2022**, *40*, 988–995. [CrossRef]

minerals

MDPI

Article

Hierarchical Intelligent Control Method for Mineral Particle Size Based on Machine Learning

Guobin Zou [1,2,3,*], Junwu Zhou [1,2], Tao Song [2,3], Jiawei Yang [2,3] and Kang Li [2,3]

[1] College of Information Science and Engineering, Northeastern University, Shenyang 110819, China
[2] State Key Laboratory of Intelligent Optimized Manufacturing in Mining & Metallurgy Process, Beijing 102628, China
[3] BGRIMM Technology Group, Beijing 102628, China
* Correspondence: zou_gb@bgrimm.com; Tel.: +86-10-5906-9777; Fax: +86-10-5906-9797

Abstract: Mineral particle size is an important parameter in the mineral beneficiation process. In industrial processes, the grinding process produces pulp with qualified particle size for subsequent flotation processes. In this paper, a hierarchical intelligent control method for mineral particle size based on machine learning is proposed. In the machine learning layer, artificial intelligence technologies such as long and short memory neural networks (LSTM) and convolution neural networks (CNN) are used to solve the multi-source ore blending prediction and intelligent classification of dry and rainy season conditions, and then the ore-feeding intelligent expert control system and grinding process intelligent expert system are used to coordinate the production of semi-autogenous mill and Ball mill and Hydrocyclone (SAB) process and intelligently adjust the control parameters of DCS layer. This paper presents the practical application of the method in the SAB production process of an international mine to realize automation and intelligence. The process throughput is increased by 6.05%, the power consumption is reduced by 7.25%, and the annual economic benefit has been significantly improved.

Keywords: machine learning; mineral particle size; hierarchical intelligent control; LSTM; CNN

check for updates

Citation: Zou, G.; Zhou, J.; Song, T.; Yang, J.; Li, K. Hierarchical Intelligent Control Method for Mineral Particle Size Based on Machine Learning. *Minerals* **2023**, *13*, 1143. https://doi.org/10.3390/min13091143

Academic Editors: Liuyi Ren, Wencheng Xia, Wei Xiao, Siyuan Yang and Dave Deglon

Received: 21 June 2023
Revised: 21 August 2023
Accepted: 24 August 2023
Published: 30 August 2023

1. Introduction

With the continuous depletion of resources and the increasing pressure of environmental protection and production costs, the production mode of mining enterprises is changing from an extensive production mode driven by human experience to an intelligent production mode driven by knowledge and data. The popularization of basic industrial process automation technology and the rapid development of big data and intelligent control technology also provide basic conditions and technical support for the intelligent operation of mineral processing [1,2]. More and more production problems that are difficult to solve by traditional methods need to be solved by artificial intelligence methods such as machine learning [3,4].

In recent years, with the improvement of computer performance, the rise of intelligent computing, and the development of AI algorithms such as machine learning and intelligent perception, machine learning technology has become a universal technology that can be used in all fields [5]. Machine learning, big data, expert systems, industrial internet, and other artificial intelligence have achieved practical applications from theoretical research, and the integration of applications in industrial scenarios shows great potential [6–10]. The industrial artificial intelligence technology represented by machine learning and intelligent perception has made breakthroughs and has been successfully applied in many fields, especially the image processing technology based on deep convolution networks, which has been increasingly applied in industrial process control, fault detection, key parameter acquisition of complex industrial systems [11–18] and other fields. Mou et al. introduced the generation antagonism network (GAN) for soft sensor modeling, established an innovative hybrid mechanism based on GAN and a data-driven soft sensor framework, and evaluated the effectiveness of the method in the industrial case of predicting the thermal deformation of the air preheater rotor of power plant boilers [19]. Wensi Ke et al. developed an LSTM-based deep neural network structure as a soft sensing method with strong nonlinearity and dynamics in the processing process and verified the effectiveness of the improved modeling method through the benchmark test of the sulfur recovery unit [20]. Yan proposed a Bayesian network (BN) based modeling and operational adjustment method is investigated [21].

Africa is rich in copper, cobalt, and other mineral resources. Because most of them are open-pit mining, the climate characteristics in the dry season and rainy season are obviously different, leading to large fluctuations in raw material properties, such as particle size and mineral composition. The existing control methods based on the determination of models and parameters have problems such as large parameter drift and weak applicability, which pose challenges to the production control of the SAB process. This is also a common problem facing the resource development of the African continent.

This paper focuses on the control problems for particle size in the SAB process of a copper concentrator in Africa. The layered intelligent control method based on machine learning is used to solve the above control problems. In the machine-learning layer, the long short memory neural network (LSTM), convolutional neural network (CNN), and other artificial intelligence technologies are used to solve the multi-source ore blending prediction and the intelligent classification of working conditions in dry and rainy seasons, Then, the intelligent expert control system for ore feeding and the intelligent expert system for semi-autogenous ball mill are used to coordinate the control of SAB process production and intelligently adjust the control parameters of DCS layer. This method has successfully improved SAB process production indicators and increased economic benefits.

2. SAB Process

2.1. SAB Process in Africa

Plant M is located in Africa, with a daily processing capacity of 10,000 tons. The SAB process is adopted for ore grinding and classification. As shown in Figure 1, The coarse ore pile is transported to the semi-autogenous mill (SAG mill). SAG mill ore discharge is screened by drum screen, and the hard stones on the screen are transported to the SAG mill by belt. The undersized slurry enters the grinding pump tank and is pumped to one group of hydrocyclones. The hydrocyclone grit returns to the ball mill for grinding. The ball mill discharge enters the grinding pump tank, and the hydrocyclone group finally forms a closed circuit. The cyclone overflows into the flotation operation. The fineness of the grinding product is −0.074 mm, accounting for 66%, and the pulp concentration is 32%, entering the downstream production process. The stability of the SAB process product production index seriously affects the production of downstream production processes and even affects the production efficiency of the entire concentrator. SAB process urgently needs to adopt automation and intelligent technology to control production stably.

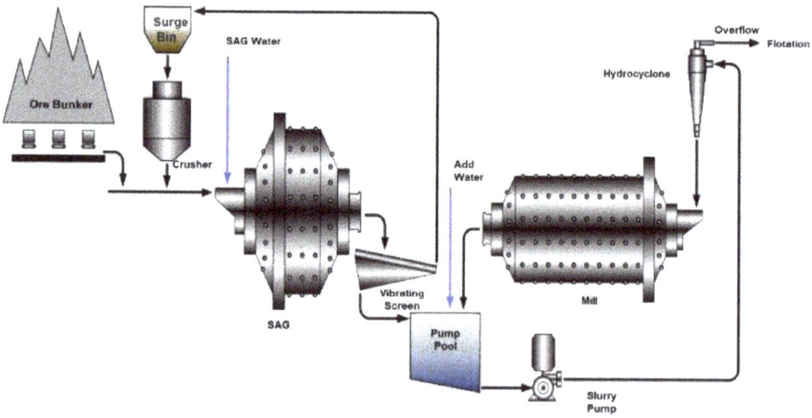

Figure 1. SAB process flowchart of Plant M.

2.2. Analysis of Operation Problems

For the SAB process of Plant M, due to the variable operating conditions of SAG grinding equipment, the operating parameters, such as power and axial pressure, vary greatly with different operating conditions. During the process operation, it is necessary to make targeted adjustments according to the change of working conditions in time to ensure that the main equipment operates in the best state. The operator is inexperienced, so it is difficult for the operator to accurately judge the operation status of the grinding production process, and cannot make correct production decisions and operate at the correct time point in time, thus affecting the overall operation efficiency and stability of the process. Different experiences and habits of different production operators will also lead to fluctuations in process indicators and equipment status between different production shifts, affecting the overall stability of the concentrator production. Due to the complex source, the raw ore obtained in the mining process is a mixture of copper sulfide and copper oxide. The mixing ratio of the two changes rapidly, resulting in large changes in hardness, lump powder ratio, and grindability of raw ore. The local climate conditions lead to obvious differences between the precipitation in the dry season and rainy season, and the open-air configuration of the grinding material pile leads to a large fluctuation of the raw ore water content with the seasons. If the ore-feeding fluctuation is not controlled, it will bring great disturbance to the system operation, making the grinding and classification process and grinding product quality fluctuate in a large range.

Due to the above characteristics, especially the problems caused by multi-source ore blending, such as the fluctuation of ore properties and the change of working conditions in dry and rainy seasons, which are difficult to solve by traditional control methods. It is also difficult to operate large equipment such as the SAG mill/ball mill stably in Plant M. As shown in Figure 2, the throughput of the SAG mill changes greatly, the main motor current of the SAG mill fluctuates frequently, and the main bearing pressure changes constantly, reflecting that the production of the SAG mill is difficult to stabilize. Production indicators such as particle size distribution and concentration of mill product will fluctuate violently, seriously affecting the production of downstream processes.

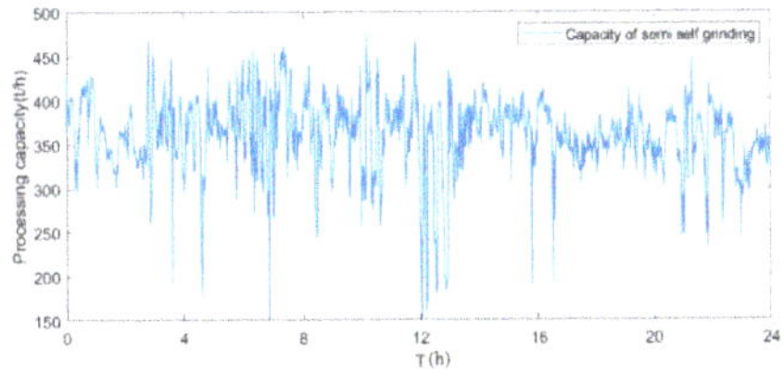

(**a**)

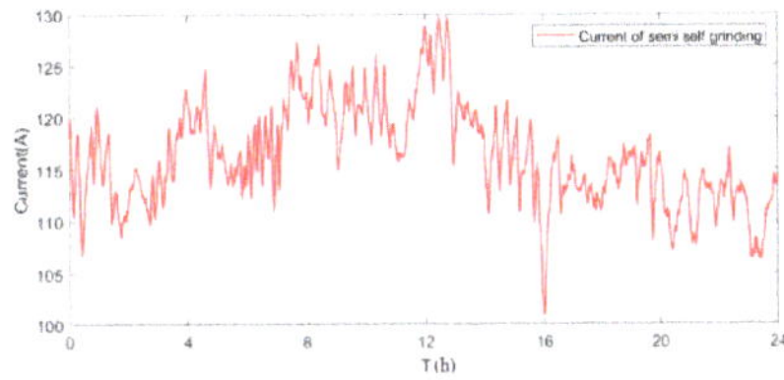

(**b**)

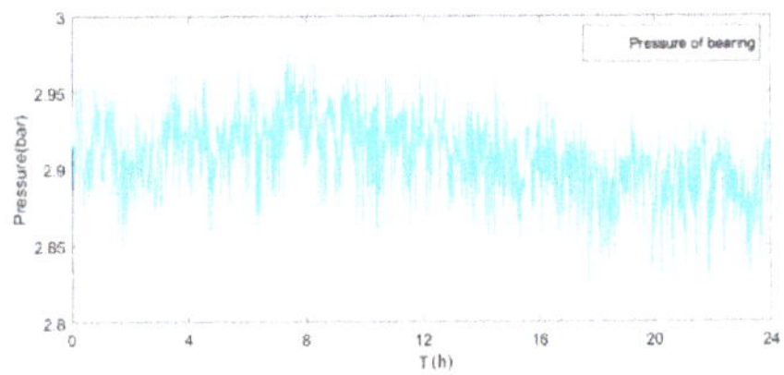

(**c**)

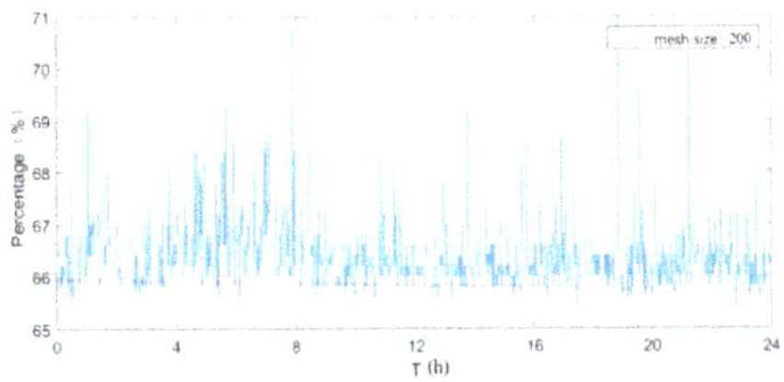

(**d**)

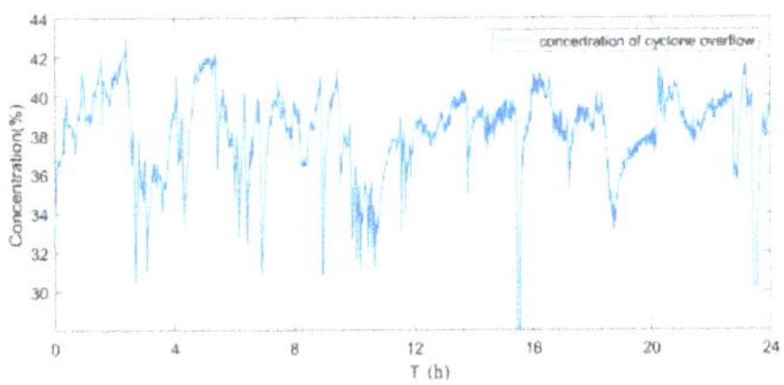

(**e**)

Figure 2. Difficulty in stabilizing the industrial grinding process. (**a**) Throughput of SAG Mill. (**b**) Current of SAG mill. (**c**) Pressure of bearing. (**d**) Percentage of mesh size (−200). (**e**) Concentration of cyclone overflow.

3. SAB Process Hierarchical Intelligent Control Method

3.1. Hierarchical Intelligent Control Structure

Machine learning is a research hotspot of industrial artificial intelligence. Its theory and method have been widely used to solve complex problems in engineering applications and scientific fields. Aiming at the problems that are difficult to solve by traditional control methods, such as the change of ore property of this process and the different working conditions in dry and rainy seasons, a layered intelligent control method of the SAB process based on machine learning is proposed, as shown in Figure 3.

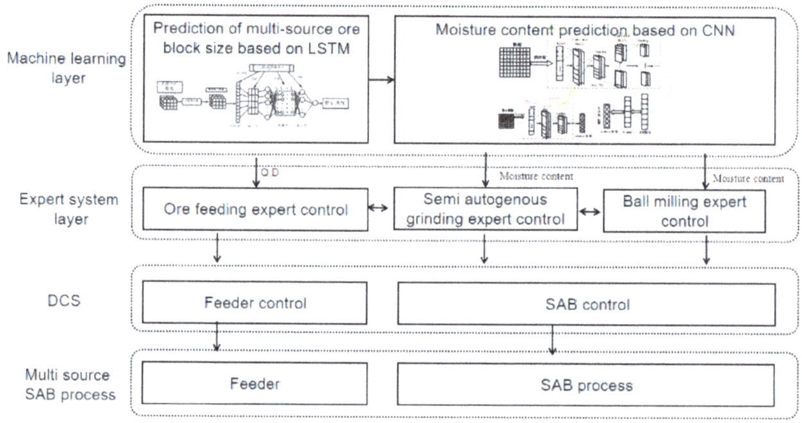

Figure 3. Hierarchical intelligent control of SAB process based on machine learning.

In the machine-learning layer, the subsequent semi-autogenous grinding and ball milling processes fluctuate dramatically due to the change of ore-feeding properties due to multi-source ore blending. In this paper, the LSTM-based semi-autogenous grinding ore-feeding block size prediction method is used to provide block size prediction information for intelligent expert control of ore. As the dry and rainy seasons lead to the change of mineral moisture, which affects the control parameters of semi-autogenous grinding and ball milling, the CNN-based intelligent classification of dry and rainy season working conditions is adopted to predict the classification of current working conditions online, adjust the current working condition parameters intelligently, and provide them to the semi-autogenous grinding expert control system and ball milling expert control system.

In the expert system layer, the intelligent expert control of ore feeding adopts the expert control method based on fuzzy rules. According to the prediction information of mineral property, the system intelligently adjusts the ore-feeding block size and ore quantity. The semi-autogenous grinding expert control and ball milling expert control adopt the CBR expert control method based on real-time compensation. According to the working condition parameters intelligently adjusted, the SAG mill and ball mill loads are intelligently coordinated, and the control parameters of the DCS layer are adjusted.

3.2. Machine Learning Layer

3.2.1. LSTM-Based Prediction Method for Feeding Lump Size of SAG

The precise control of ore feeding is the premise of efficient and stable operation of SAG and also the basis of the whole grinding intelligent control. For semi-autogenous grinding ore feeding, its specific control objectives include the stability of the ore-feeding amount and the stability of the ore-feeding lump size. Due to ore blending at multiple sources, the rock size in feed ore to the SAG mill will fluctuate, affecting the production stability of subsequent ball milling equipment. If the fragmentation of SAG mill input ore can be predicted according to the production data, the fragmentation change can be predicted in time when the working conditions are shifted due to multi-source ore, and the expert control system of the SAG mill can be guided to quickly adjust and reach the new set value, which is conducive to the stability of the whole system.

There are four ore feeders at the bottom of the raw ore bin. The location of each feeder is different. The corresponding storage height and ore fragmentation from feeders are also different and will change with the production process. The ore fragmentation shows obvious time-varying and nonlinear features. According to experts' experience and knowledge, the variables that can be detected online and have a great impact on the block size are selected, as shown in Table 1.

Table 1. Online detection variables of SAG.

No.	Variable Name
1	1# Belt frequency
2	2# Belt frequency
3	3# Belt frequency
4	4# Belt frequency
5	Feedrate of SAG mill
6	Current of SAG mill
7	Water feedrate of SAG mill
8	Pressure of SAG mill bearing
9	SAG mill 2# Belt current

Through the combination and verification of manual belt sampling and screening + video image manual marking division, the actual ore lump size on the semi-autogenous grinding feeding belt in a period of time can be quickly obtained. The long short memory neural network is used to train the prediction model of the ore block size for semi-autogenous grinding. The LSTM training network used in this paper is shown in Figure 4.

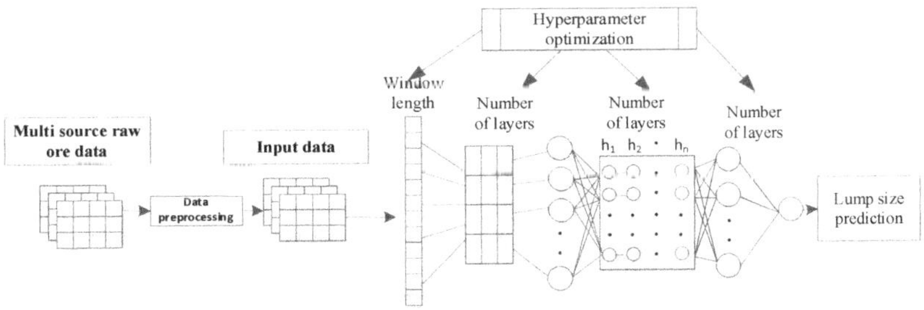

Figure 4. Prediction of grinding fragmentation based on LSTM.

In the process of LSTM modeling, the problem of overfitting often occurs. The model network structure is too complex, which exceeds the actual problem. It performs well in the training set but performs poorly in the test set and has poor generalization performance. Therefore, Dropout is used to reduce the overfitting phenomenon when training the network. Adam optimizer is selected for updating model parameters. The RMSE and ARGE are calculated by the formulas in [22–24].

The LSTM network is a deep neural network with many superparameters for predicting the feeding lump size of the SAG mill. Since all neural networks have an input layer and an output layer, the complexity of the deep learning model mainly depends on the number of hidden neurons and the number of neurons in each layer, which is the main superparameter of the deep learning model. In addition to the network structure, the LSTM network model also has this time window size, which is more parameters and more complex to optimize than the general deep neural network. The superparameter learning method with the following structure is used to optimize the network parameters.

As shown in Figure 5. SQP algorithm is used to optimize the superparameters of the prediction model for the feed size of the SAG mill. The optimization problem is expressed as follows [25]:

$$\begin{cases} \min f(x) \\ subject\ to\ g_i(x) \leq 0\ (i = 1, 2, \ldots, m_p) \\ \qquad\qquad g_i(x) = 0\ (i = m_p + 1, \ldots, m) \end{cases} \tag{1}$$

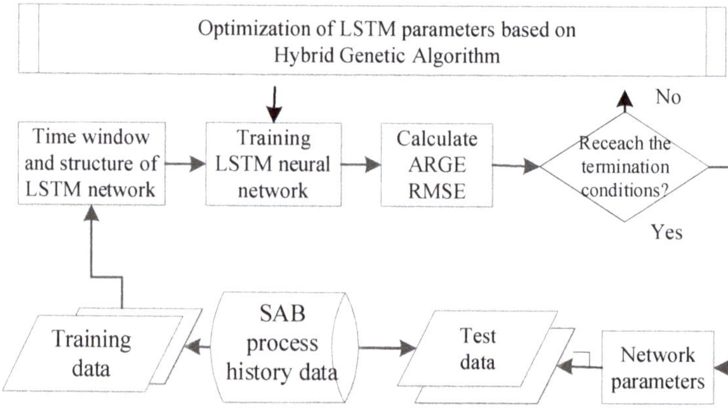

Figure 5. Optimization of LSTM parameters based on hybrid genetic algorithm.

For the soft-sensing detection of the feeding lump size of the SAG mill, GA and HGA are used to optimize the hidden layer of the LSTM network, the unit number of the hidden layer, and the Dropout probability. In order to ensure the validity of the comparison of model performance results, the superparameter settings that do not involve optimization should be the same. In the HGA optimization experiment, the modeling based on GA-LSTM is first carried out. The initial population size of the genetic algorithm [1] is set to 30, the crossover rate is 0.5, the mutation rate is 0.1, and the total number of iterations is set to 20. When the genetic algorithm is used to optimize the super parameters of the LSTM network, the optimal fitness function of the optimal chromosome in the population, namely the RMSE value of the LSTM network on the test set, gradually converges with the increase of iteration times. The lump size prediction effect of SAG based on LSTM is shown in Figure 6.

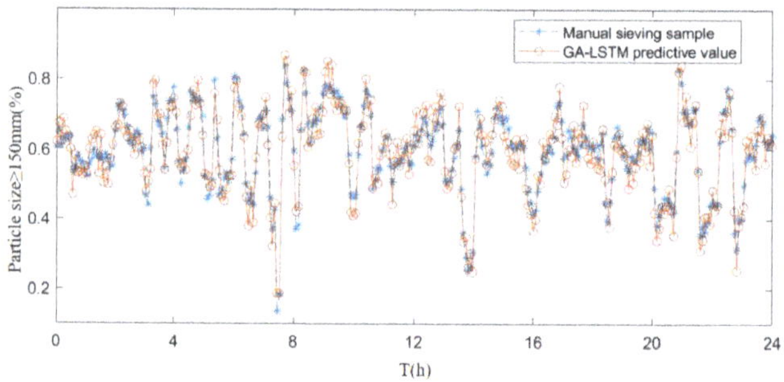

Figure 6. Lump size prediction of SAG mill feed.

3.2.2. Classification of Moisture Content in Dry and Rainy Seasons Based on CNN

Due to the local climate conditions, the precipitation in the dry season is obviously different from that in the rainy season, while the open pit configuration of the grinding material pile results in a large fluctuation of the raw ore water content with different seasons. In the dry season, the water content of raw ore is about 8%–10%, while in the rainy season, when the precipitation is large, the water content of raw ore can reach about 15%. This brings great interference to the grinding production control and affects the control effect of the optimal control system in the rainy season. In order to improve the adaptability of the control system, it is necessary to realize the adaptive control algorithm parameters.

A total of 24,500 pieces of data covering 24 months, as shown in Table 2, including the dry season and rainy season. Moreover, 15,000 pieces of data were used for modeling training, 5000 pieces of data were used for modeling testing, and 4500 pieces of data were used for modeling validation.

Table 2. Data variables for CNN training.

No.	Variable Name
1	Mesh size (Video extraction)
2	Feedrate of SAG mill
3	Water feedrate of SAG mill
4	Current of SAG mill
5	Bearing pressure of SAG mill
6	Current of ball mill
7	Bearing pressure of ball mill
8	Water feedrate to sump
9	Pressure of cyclone
10	Flow of cyclone
11	Concentration of cyclone
12	Particle size of cyclone overflow

A CNN network using VGG-16 structure [26,27], with five segments connected in series by multiple 3 × 3 convolutional kernels. The maximum pooling layer with a size of 2 × 2 is connected after each convolutional segment. The network scales the feature map through the pooling layer, with three fully connected layers and one softmax layer connected at the end [28]. The network structure in Figure 7 is used to classify the data during the dry and rainy seasons, and the classification results are combined with manual labeling, as shown in Figure 8.

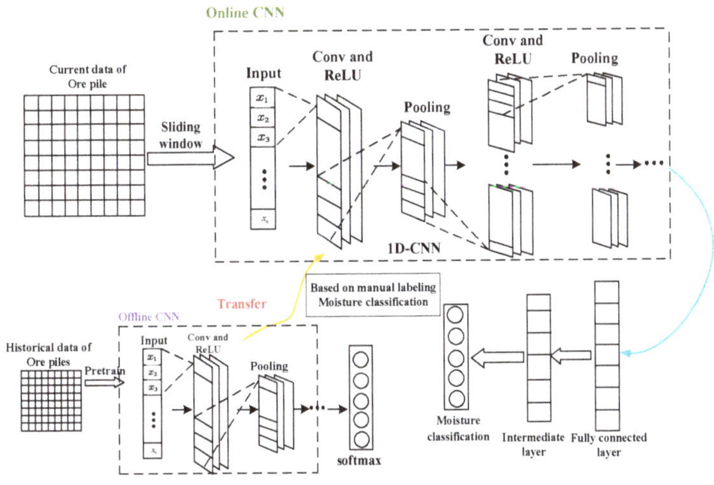

Figure 7. Classification of moisture content in dry and rainy seasons based on CNN.

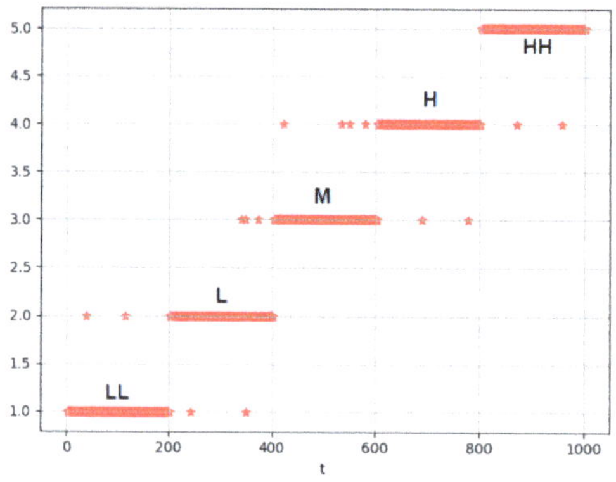

Figure 8. Results of classification of moisture content in dry and rainy seasons based on CNN.

3.3. Expert Control Layer

3.3.1. Feeding Expert Control

The precise control of ore feeding is the premise of efficient and stable operation of the SAG mill as shown in the Table 3 and also the basis of the whole grinding intelligent control. For semi-autogenous grinding ore feeding, its specific control objectives include two aspects: the stability of the ore-feeding amount and the stability of ore-feeding particle size distribution. In detail, one is to maintain the stability of mineral feeding materials under normal working conditions, and another is to quickly adjust and reach the new set value when working conditions migrate, which is conducive to the stability of the entire system.

Table 3. Online detection variables of SAG mill.

IF	THEN
LL	Parameter Group 1 of Expert control layer
L	Parameter Group 2 of Expert control layer
M	Parameter Group 3 of Expert control layer
H	Parameter Group 4 of Expert control layer
HH	Parameter Group 5 of Expert control layer

Difficulties of intelligent feeding include large system lag, sometimes variable storage and measuring tools in the silo, differences among feeders, and sudden occurrence of various abnormal conditions on-site. As for Plant M, there are four feeders at the bottom of the raw ore bin. The location of each feeder is different, and its corresponding storage height and ore block size are different and change with the production process. The ore-feeding capacity and ore block size show obvious time variation and difference. The distance between the feeder and the belt scale (measuring mechanism) is about 200 m, with a delay of about 2 min. In the rainy season, the feeder downport is easy to be blocked. At this time, the operators must stop the corresponding feeder immediately and start a new feeder to ensure the continuity of production. In actual production, materials in raw ore are mixed with various ores with different copper grades and particle sizes according to the feedback scheduling of the flotation process. Aiming at the actual problems on-site, the feeding expert control system compensates for the dynamic changes of the feeding process through the LSTM-based model of the feeding process and conducts real-time control on the amount and fragmentation of SAG mill feed to ensure the stability of grinding production as shown in Figure 9.

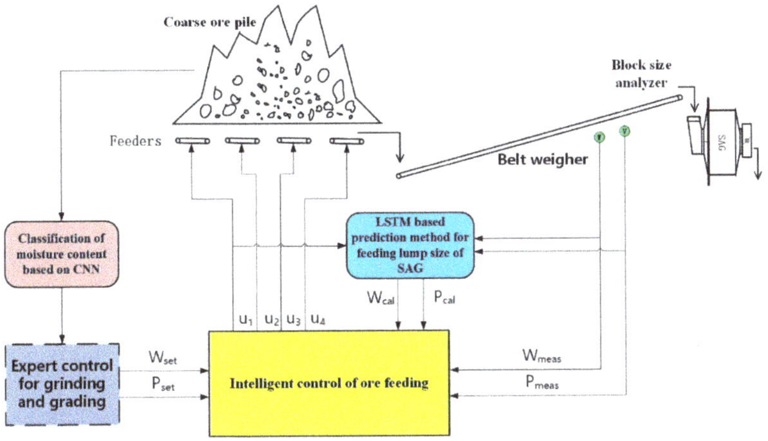

Figure 9. Intelligent control of ore-feeding plus compensation module.

If the ore-feeding capacity of the ith feeder ($1 \leq i \leq 4$) is k_i, the proportion of large blocks is p_i, its frequency value is u_i, the prediction value of ore-feeding amount is W_{cal}, and the prediction value of the proportion of large blocks is P_{cal}, then:

$$\begin{cases} \sum_{i=1}^{4} k_i \cdot u_i = W_{cal} \\ \sum_{i=1}^{4} p_i \cdot k_i \cdot u_i = P_{cal} \cdot W_{cal} \\ u_{min} \leq u_i \leq u_{max} \end{cases} \tag{2}$$

When there is a deviation between the actual value W_{means} of the ore-feeding amount measured by the ore-feeding belt scale and the P_{meas} measured by the ore lump analyzer and the set value W_{set} of the ore-feeding amount and the set value P_{set} of the ore-feeding lump proportion given by the grinding and classification expert control, the adjustment amount of the feeder frequency can be calculated by the increment method:

$$\begin{cases} \sum_{i=1}^{4} k_i \cdot \Delta u_i = W_{meas} - W_{set} \\ \sum_{i=1}^{4} p_i \cdot k_i \cdot \Delta u_i = P_{meas} \cdot W_{meas} - P_{set} \cdot W_{set} \\ -u_{max} \leq \Delta u_i \leq u_{max} \\ u_{min} \leq u_i + \Delta u_i \leq u_{max} \end{cases} \tag{3}$$

In the actual control, according to the coarse and fine classification of the ore feeder's blanking block size, the frequency of one coarse and one fine pair of ore feeders can be adjusted each time, so the above equation can be degenerated into a two-dimensional linear equation, which can be solved quickly. When the ore-feeding frequency is lower than the lower limit or higher than the upper limit, poll other ore-feeding port matching schemes or turn on/off the feeder.

When it is necessary to open or close the feeder, in order to ensure that the belt materials are not stacked and empty, the physical position of the ore-feeding mouth and the speed of the transmission belt are measured, and the switching sequence and time of the feeder at each ore-feeding mouth are precisely controlled, so as to realize the continuous and stable transmission of the mineral material flow on the ore-feeding belt.

In view of the large delay (2 min) of feeding machine blanking and belt weigher measurement value, a large interval sampling control is adopted. The cycle of the controller is greater than the maximum lag time of each feeder with state change in the last control cycle plus the maximum adjustment time of the feeder (5 s~10 s). Because the control effect of the regulator output is unknown within the lag time. The basic idea of large interval sampling is to wait for a period of time after one adjustment until the end of the adjustment process, and the measurement result of the belt scale can fully reflect the real ore-feeding amount before the next adjustment so as to avoid false adjustment due to misjudgment, resulting in closed-loop system oscillation or instability. Through large interval sampling control, it can realize stable ore feeding when switching feeder and adjust in place as soon as possible in case of error.

The effect after the expert control system is shown in Figure 10. The fluctuation range of fragmentation is reduced, and the production process is more stable, meeting the production index requirements.

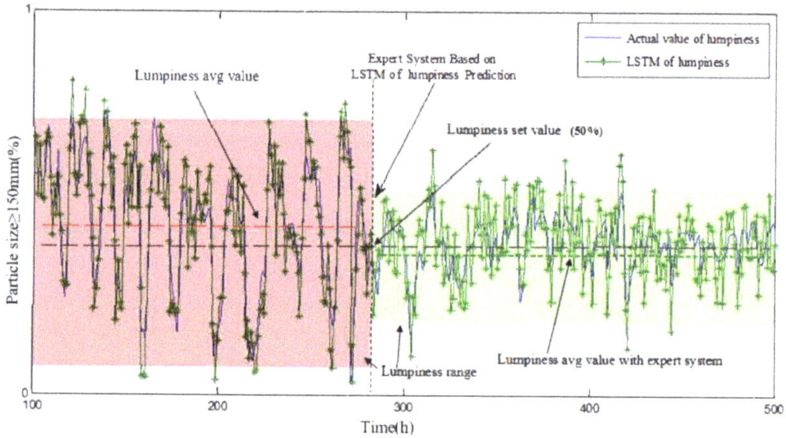

Figure 10. The stability of feeding lump size of the SAG controlled by experts.

3.3.2. CBR Expert Control Method

According to the characteristics of the new problem, the case-based reasoning process first compares the relevance of the historical cases in the case base and then selects one or several cases that meet a certain degree of similarity from the case base. Similar case solutions are modified to be the solutions to new problems. And through the effect of solving the problem, it is decided whether to store the case of the new problem into the case base so as to solve the following problems. Some edge conditions cause similar cases to be found in the case retrieval process, or the case matching similarity is low. So, these cases need to be translated and interpreted through the previously set production rule base and solved through the pre-chain rules, which are applied to the scene, solving the production misoperation when the case matching similarity is not high.

Because there is a certain difference between the retrieved case and the current working condition, there is also a certain difference between the case solutions. Therefore, the case solution cannot be directly used as the solution to the current working condition. In order to further improve the accuracy of the model, the RBF neural network method is used to establish an incremental compensation model to compensate for the retrieved case solution.

4. Application Effect and Analysis

4.1. Intelligent Control System

As shown in Figure 11. The Grinding Process Master software(BPM-G v1.0) has been developed by using the above technologies. In this SAB process, the closed-loop control and intelligent optimization of operating variables such as ore feed rate, water feed rate, and sand pump frequency have been realized, ensuring the safe, stable, and efficient operation of the entire production process. At present, the SAB process has been in actual operation for more than one year.

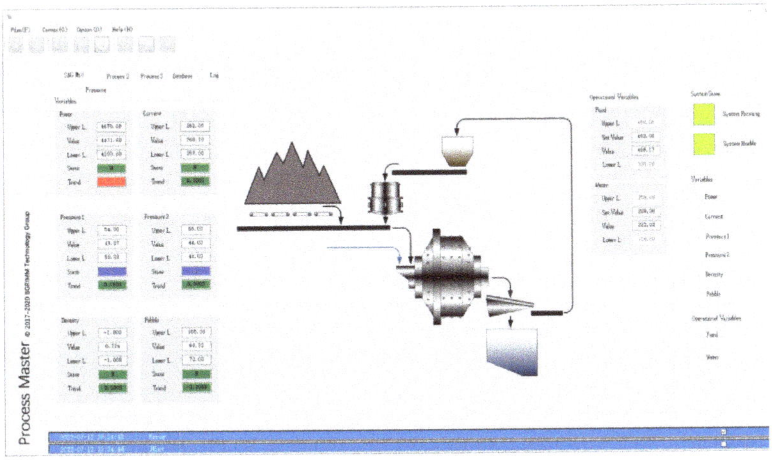

Figure 11. The stability of feeding lump size of the SAB controlled by experts.

4.2. Application Effect

After the application of the grinding optimization control system, the closed-loop automatic control of the grinding and classification process is realized, the intensity of manual operation is reduced, the dependence on operating experience is reduced, and labor productivity is improved. Based on stabilizing the production process and reducing the abnormal disturbance in the production process, the system processing capacity is increased, and the production unit consumption is reduced.

Through real-time intelligent diagnosis of production process operation status, the system timely and reasonably adjusts production operation variables to ensure the stability of production equipment operation status and production process parameters. After the system is put into use, the fluctuations of key production process indicators are reduced by more than 20%. The grinding expert system stabilizes the operation parameters of each piece of equipment in the SAB process within a reasonable range according to different ore sources. The stability of the process operation ensures the stability of the grinding and classification of product quality. The grinding particle size is more concentrated in the optimal mesh size area, as shown in Figure 12.

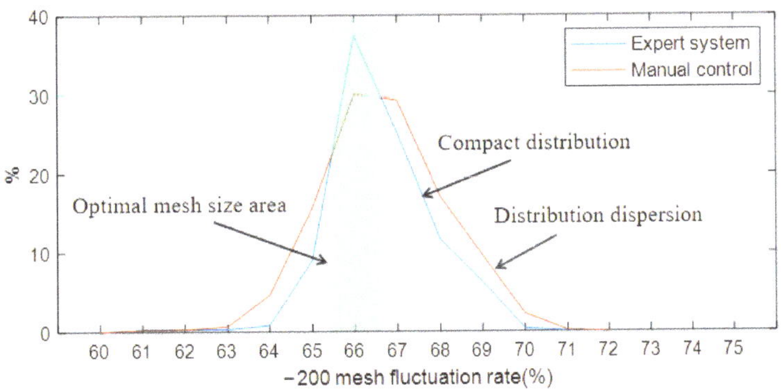

Figure 12. Particle size distribution statistics of grinding products.

As shown in Table 4, after the system is put into use, the grinding expert system can sensitively capture the changing trend of the key operating parameters of the SAG mill, adjust the efficiency in time when the working conditions change, and ensure the SAG mill to operate in a reasonable load state range to the greatest extent. On the premise of stabilizing the working conditions and ensuring the process indicators, the production efficiency of the process is improved. The average processing capacity of grinding production increased by 6.05%.

Table 4. Statistical comparison of production parameters on 6 months data.

Variable Name	Unit	Before	After	Improve
Throughput fluctuation rate	%	33.81%	17.72%	−47.6%
Current fluctuation rate of main motor	%	10.39%	5.89%	−43.3%
Bearing pressure fluctuation rate	%	10.78%	7.01%	−35.0%
−200 mesh fluctuation rate	%	6.72%	3.45%	−48.7%
Concentration fluctuation rate	%	12.93%	8.39%	−35.1%

By stabilizing the operation state of the production process, the system ensures that the process always operates in a state of high economic benefits under different ore properties, reduces the unit production consumption, and reduces the wear of the lining plate. According to the statistical data of the grinding system, including process parameter indicators such as unit efficiency and energy consumption (power consumption, ball consumption), the power consumption per ton of mill ore decreased by 7.25% on average, increasing the total amount of ore processing within the service life of a single set of liner plates.

5. Conclusions

Aiming at the control problems of mineral particle size in the SABC grinding process, this paper proposes an intelligent expert control method of the SABC process based on a hybrid model. A hybrid model of SVR and mechanism model is adopted to realize the online perception of overflow fineness. The adaptability of the expert rule base is improved by intelligent correction of uncertainty, and then combined with the experience of field operation experts, intelligent optimization control of overflow fineness and unit power consumption is realized.

Author Contributions: Data curation, G.Z. and J.Z.; formal analysis, G.Z. and J.Y.; funding acquisition, K.L.; methodology, G.Z. and K.L.; software, T.S.; writing—original draft, G.Z. All authors have read and agreed to the published version of the manuscript.

Funding: The support from the following foundations: the National Key R&D Program of China [grant number: 2021YFC2902700] and the National Natural Science Foundation of China [grant number: 62273078].

Data Availability Statement: Not applicable.

Conflicts of Interest: The authors declare no conflict of interest.

References

1. Zou, G.; Zhou, J.; Li, K.; Zhao, H. An HGA-LSTM-Based Intelligent Model for Ore Pulp Density in the Hydrometallurgical Process. *Materials* **2022**, *15*, 7586. [CrossRef]
2. Fan, C.; Ren, L.; Zhang, Y.; Bao, S. Grinding effect of sodium silicate on muscovite and its mechanism analysis. *Miner. Eng.* **2023**, *199*, 108106. [CrossRef]
3. Sadr, H.; Pedram, M.M.; Teshnehlab, M. Multi-View Deep Network: A Deep Model Based on Learning Features From Heterogeneous Neural Networks for Sentiment Analysis. *IEEE Access* **2020**, *8*, 86984–86997. [CrossRef]
4. Tian, Y.H. Artificial Intelligence Image Recognition Method Based on Convolutional Neural Network Algorithm. *IEEE Access* **2020**, *8*, 125731–125744. [CrossRef]
5. Jiang, Y.; Yin, S.; Dong, J.; Kaynak, O. A review on soft sensors for monitoring, control, and optimization of industrial processes. *IEEE Sens. J.* **2020**, *21*, 12868–12881. [CrossRef]
6. Kadlec, P.; Gabrys, B.; Strandt, S. Data-driven Soft Sensors in the process industry. *Comput. Chem. Eng.* **2009**, *33*, 795–814. [CrossRef]

7. Lu, Y.; Lu, G.M.; Lin, R. SRGC-Nets: Sparse Repeated Group Convolutional Neural Networks. *IEEE Trans. Neural Netw. Learn. Syst.* **2019**, *31*, 2889–2902. [CrossRef] [PubMed]
8. Li, K.; Daniels, J.; Liu, C.; Herrero, P.; Georgiou, P. Convolutional Recurrent Neural Networks for Glucose Prediction. *IEEE J. Biomed. Health Inform.* **2020**, *24*, 603–613. [CrossRef] [PubMed]
9. Dai, J.; Chen, N.; Yuan, X.; Gui, W.; Luo, L. Temperature prediction for roller kiln based on hybrid first-principle model and data-driven MW-DLWKPCR model. *ISA Trans.* **2020**, *98*, 403–417. [CrossRef]
10. Kim, B.; Yuvaraj, N.; Preethaa, K.R.S.; Pandian, R.A. Surface crack detection using deep learning with shallow CNN architecture for enhanced computation. *Neural Comput. Appl.* **2021**, *33*, 9289–9305. [CrossRef]
11. Zhou, P.; Gao, B.; Wang, S.; Chai, T. Identification of Abnormal Conditions for Fused Magnesium Melting Process Based on Deep Learning and Multi-source Information Fusion. *IEEE Trans. Ind. Electron.* **2021**, *99*, 1.
12. Zhang, H.; Tang, Z.; Xie, Y.; Chen, Q.; Gao, X.; Gui, W. Feature Reconstruction-Regression Network: A Light-Weight Deep Neural Network for Performance Monitoring in the Froth Flotation. *IEEE Trans. Ind. Inform.* **2020**, *17*, 8406–8417. [CrossRef]
13. Zhang, J.; Tang, Z.; Xie, Y.; Chen, Q.; Ai, M.; Gui, W. Timed key-value memory network for flotation reagent control. *Control Eng. Pract.* **2020**, *98*, 104360. [CrossRef]
14. Gao, F.; Ma, F.; Wang, J.; Sun, J.; Yang, E.; Zhou, H. Semi-Supervised Generative Adversarial Nets with Multiple Generators for SAR Image Recognition. *Sensors* **2018**, *18*, 2706. [CrossRef]
15. Hinton, G.E.; Salakhutdinov, R.R. Reducing the Dimensionality of Data with Neural Networks. *Science* **2006**, *313*, 504–507. [CrossRef] [PubMed]
16. Zhou, K.; Yang, C.; Liu, J.; Xu, Q. Deep graph feature learning-based diagnosis approach for rotating machinery using multi-sensor data. *J. Intell. Manuf.* **2022**, *34*, 1965–1974. [CrossRef]
17. Zhou, X.; Wang, Q.; Zhang, R.; Yang, C. A hybrid feature selection method for production condition recognition in froth flotation with noisy labels. *Miner. Eng.* **2020**, *153*, 106201. [CrossRef]
18. Zhang, H.; Tang, Z.; Xie, Y.; Yuan, H.; Chen, Q.; Gui, W. Siamese Time Series and Difference Networks for Performance Monitoring in the Froth Flotation Process. *IEEE Trans. Ind. Inform.* **2021**, *18*, 2539–2549. [CrossRef]
19. Mou, L.; Ghamisi, P.; Zhu, X.X. Deep Recurrent Neural Networks for Hyperspectral Image Classification. *IEEE Trans. Geosci. Remote Sens.* **2017**, *55*, 3639–3655. [CrossRef]
20. Ke, W.; Huang, D.; Yang, F.; Jiang, Y. Soft sensor development and applications based on LSTM in deep neural networks. In Proceedings of the 2017 IEEE Symposium Series on Computational Intelligence (SSCI), Honolulu, HI, USA, 27 November–1 December 2017. [CrossRef]
21. Yan, H.; Wang, F.; He, D.; Zhao, L.; Wang, Q. Bayesian Network-Based Modeling and Operational Adjustment of Plantwide Flotation Industrial Process. *Ind. Eng. Chem. Res.* **2020**, *59*, 2025–2035. [CrossRef]
22. Sherstinsky, A. Fundamentals of Recurrent Neural Network (RNN) and Long Short Term Memory (LSTM) Network. *Phys. D Nonlinear Phenom.* **2020**, *404*, 132306. [CrossRef]
23. Pan, H.; Su, T.; Huang, X.; Wang, Z. LSTM-based soft sensor design for oxygen content of flue gas in coal-fired power plant. *Trans. Inst. Meas. Control* **2021**, *43*, 78–87. [CrossRef]
24. Alshwaheen, T.I.; Hau, Y.W.; Ass'Ad, N.; Abualsamen, M.M. A Novel and Reliable Framework of Patient Deterioration Prediction in Intensive Care Unit Based on Long Short-Term Memory-Recurrent Neural Network. *IEEE Access* **2020**, *9*, 3894–3918. [CrossRef]
25. Mansoornejad, B., Mostoufi, N.; Jalali-Farahani, F. A hybrid GA–SQP optimization technique for determination of kinetic parameters of hydrogenation reactions. *Comput. Chem. Eng.* **2008**, *32*, 1447–1455. [CrossRef]
26. Gu, K.; Xia, Z.F.; Qiao, J.F. Deep dual-channel neural network for image-based smoke detection. *IEEE Trans. Multimed.* **2020**, *22*, 311–323. [CrossRef]
27. Xu, G.W.; Liu, M.; Jiang, Z.F.; Huang, C.X. Online fault diagnosis method based on transfer convolutional neural networks. *IEEE Trans. Instrum. Meas.* **2019**, *69*, 509–520. [CrossRef]
28. Fang, C.; He, D.; Li, K.; Liu, Y.; Wang, F. Image-based thickener mud layer height prediction with attention mechanism-based CNN. *ISA Trans.* **2022**, *128*, 677–689. [CrossRef] [PubMed]

MDPI
St. Alban-Anlage 66
4052 Basel
Switzerland
www.mdpi.com

Minerals Editorial Office
E-mail: minerals@mdpi.com
www.mdpi.com/journal/minerals

www.ingramcontent.com/pod-product-compliance
Lightning Source LLC
Chambersburg PA
CBHW062032220926
51850CB00010B/2524
* 9 7 8 3 7 2 5 8 1 1 3 6 6 *